RAIN GARDENS

RAIN GARDENS

Managing water sustainably in the garden and designed landscape

Nigel Dunnett and Andy Clayden

Timber Press
Portland • London

The Aztec Business Park, Bristol, UK, makes use of ponds and naturalistic planting as part of its landscape. Not only does this make for an interesting and stimulating work environment, but swans have come in to nest alongside the office buildings.

Photographs by Nigel Dunnett and Andy Clayden unless otherwise stated. Illustrations by Andy Clayden.

Published in 2007 by Timber Press, Inc.

The Haseltine Building
133 S.W. Second Avenue, Suite 450
Portland, Oregon 97204-3527
www.timberpress.com

2 The Quadrant
135 Salusbury Road
London NW6 6RJ
www.timberpress.co.uk

Third printing 2008

Design by Dick Malt
Printed in China

Library of Congress Cataloging-in-Publication Data

Dunnett, Nigel.
  Rain gardens : sustainable rainwater management for the garden and designed landscape / Nigel Dunnett & Andy Clayden.
      p. cm.
  Includes bibliographical references and index.
  ISBN-13: 978-0-88192-826-6
  1. Landscape gardening--Water conservation. 2. Water harvesting. 3. Gardens--Design. 4. Plants, Ornamental--Varieties. I. Clayden, Andy.
  II. Title.
  SB475.83.D86 2007
  635.9'5--dc22
                                2006035032
A catalogue record for this book is also available from the British Library.

CONTENTS

ACKNOWLEDGEMENTS

The inspiration behind this book lies firmly at the feet of one man: Tom Liptan, Stormwater Coordinator for the City of Portland, Oregon. Tom has been a tireless advocate of the need to bring nature and natural processes to the fore in the way that we plan for healthy cities. His work in Portland is a testament to that dedication and achieves environmental function and benefit in a way that is creative and aesthetically beautiful. That is the philosophy behind this book.

We also wish to acknowledge the generosity of others in supporting the book, in particular for donating pictures and advice. We are particularly thankful to Erin Middleton for her efforts in supplying details and pictures of the rain gardens created in Portland by Urban Water Works; John Little, of the Grass Roof Company, London; and Vaughn Wascovich and Vicky Ranney of Prairie Crossing for permission to use pictures of their environmentally-sensitive residential scheme.

Finally, we wish to thank Anna Mumford, our editor at Timber Press, for her enthusiasm and support throughout the writing of this book, and for the friendliness of all at Timber Press.

Our demand for water is growing dramatically at a time when in many areas of the world its supply can no longer be guaranteed. Even in the comparatively short time of writing *Rain Gardens* there has been an escalation in media coverage on issues relating to global warming and the increased probability of serious summer drought. Hosepipe bans are now common in the UK and in London there is a real prospect that drinking water will be supplied from street standpipes in exceptionally dry summers. Rain gardens represent a radical change in our thinking about how we manage rainfall. By capturing and using rainwater we not only significantly reduce our demand for treated water, we also have an opportunity to rethink how we design and manage public and private open spaces in order to improve their environmental and aesthetic quality.

Rain Gardens explores the many different ways in which rainfall can be captured from buildings and sealed surfaces and then stored and released within the landscape. Technical details explain how different design elements are constructed, while case studies from the United States and northern Europe show exactly how the concept has been applied to a range of settings, including residential developments, schools, public parks, city squares and private gardens. A key feature of each of these schemes is the enhanced role of planting and planting design. Rain gardens enable designers and those who enjoy working with plants to argue for their enhanced role in the design of our public and private spaces not just for their aesthetic qualities but also—and most importantly—for their environmental and economic contribution.

The gardens of the Generallife,
Granada, Spain, include the
classic elements of the
Paradise Garden—water,
shade, colour and scent.

INTRODUCTION

Without water we wouldn't have gardens. Water was a precious resource to the ancient cultures of the Middle East, where the concept of gardens designed purely for pleasure first originated. The original Paradise Gardens were stylized versions of the irrigation channels that brought life to parched desert lands, and the biblical Garden of Eden represents a lush green (water-fed) vision of perfection to peoples more used to arid surroundings. We are magnetically attracted to water—throughout history, pools, ponds, lakes, streams and fountains have been indispensable ornamental elements in gardens, and more recently the presence of water has been seen as essential to gardens that aim to attract wildlife as well as to provide pleasure for people. We have now come full circle, with water once again viewed as a finite and unpredictable resource, taking centre stage in the way we plan and manage our gardens and landscapes, not just because of its aesthetic and environmental value, but because its presence or absence ultimately dictates what is possible.

Our view of water is changing: it is no longer in unlimited, cheap supply, or under our control; it is instead a potentially destructive force, and certainly one that we can no longer subjugate or take for granted. With this new view has come a new way of working with water—a way that is about discovering how water might behave naturally in our surroundings, and working as hard as possible to allow and enable it to do so. This way of working is an environmentally-friendly and positive approach to tackling the seemingly increasingly frequent water-related problems that we hear about and experience, both at home and in distant countries.

These are big issues, and it may seem strange to be discussing them

Ponds and pools, both formal and informal, have been at the centre of garden making across cultures and ages.

in a book focused primarily on gardens, designed landscapes, and planting. Surely these are matters more suited to engineers, economists and politicians, and more relevant to the large-scale planning of our towns, cities and countryside? Well, to some extent, yes. But this is also an instance where small-scale actions at an individual level can have real impact. Residential areas, often with large proportions of garden space, represent significant coverage in a town or city, and collectively present important opportunities to influence our demand for water, and the way that water works in our developed environments. Residential landscapes are not exactly blameless in the debate over water usage, conservation and supply. It became apparent in the latter part of the 20th century that standard garden water management was becoming unsustainable. For example, in years of normal rainfall, landscape irrigation accounted for 43 per cent of all residential water use in the western USA, and a surprising 26 per cent in the wetter eastern USA (Thayer 1982). Much of that water is used to irrigate turf grass in dry regions, often using far more water than is needed.

Our aim in writing this book is to look into and explain how simple techniques can make a real difference to the way that water is managed in designed landscapes. Many of the approaches discussed in this book have been applied by landscape architects and architects on large-scale schemes—commercial, factory and office developments, urban public squares, or housing schemes. We will highlight some of the best of these schemes and indicate how they can be modified or adapted to suit different scales of application. Moreover we firmly believe that taking a new and fresh look at how water can be used in designed landscapes represents an exciting and fulfilling opportunity for a different approach to garden and landscape design, but also one that effectively

High demand for water in residential areas might be partially addressed by collecting rain that falls on roofs and other impermeable surfaces.

ties together buildings (whether they be a domestic house, a commercial building, or a school or college) and their surrounding environments in a way that has not been expressed so forcefully before. Overall, we aim to show how water can be the fundamental basis for the design of a landscape that is not only beautiful, but which makes a significant contribution to important environmental problems that affect us all: a basis that brings to life for people the way our landscapes work and provides a very creative and satisfying underlying philosophy for all garden and designed landscapes.

The water-sensitive landscape takes a new view of how water works in our surroundings. The pond in the grounds of this new school in Germany is fed by water that drains from the roof, paths and soil, and regularly mown grass is replaced by wildflower meadow.

The 2001 Potsdam garden festival in Germany included a children's play area. For older children there was a large rectangular pool with stepping stones, a floating platform and a jetty.

SECTION 1

Rain gardens

INTRODUCTION

This book is all about the emerging idea of the rain garden. From relatively modest beginnings in the late 1980s in the state of Maryland, USA, the concept has mushroomed into one of the fastest growing areas of interest for the development of home landscapes. Unlike many other environmental initiatives that challenge us to change our lifestyles, often at personal cost, and that usually focus on the scientific or environmental benefits of what we may be doing, rain gardens have the potential to be beautiful additions to our environment, and bring with them a host of other benefits.

As will be shown later in the book, technically a rain garden has come to mean something very specific, namely a planted depression that is designed to take all, or as much as possible, of the excess rainwater run-off from a house or other building and its associated landscape. However, we feel that the term is such an evocative one that we use a much more wide-ranging definition, which covers all the possible elements that can be used to capture, channel, divert and make the most of the natural rain and snow that falls on a property. The whole garden becomes a rain garden, and all of the individual elements that we deal with in detail are either components of it, or are small-scale rain gardens in themselves. Rain gardens are therefore about water in all its forms, still and moving, above and below ground, and the rich planting and experiential opportunities that exploiting that water can give rise to.

The book is divided into three sections. This first section takes a wider look at the environmental role of water in a garden and considers some of the key principles that underlie the rain garden approach. The second section looks in detail at the individual components of the rain garden, and considers how they can be linked together as a coherent whole, using the concept of the 'stormwater chain' as the unifying element. Section Three is a technical section comprising a rain garden plant directory. Many examples of rain gardens take a fairly piecemeal approach—siting one or two features in isolation with the result that a rather bitty landscape develops. We want to look at how a more satisfying and holistic framework can be developed. This book will appeal to garden designers and to landscape architects (and students of those subjects), enthusiastic gardeners who have a strong feeling of environmental responsibility, as well as to those who have a passion for water gardening, and to architects and urban designers of innovative solutions to water management in new residential developments, school grounds, campus landscapes or business and commercial developments.

The rain garden idea is therefore very different from the concept of xeriscaping, as practised in very dry climatic zones, such as the desert and Mediterranean climates of the arid south-western USA. This is an equally ecological approach, using plantings that require little or no irrigation for successful growth, as an alternative to the traditional water-hungry landscape. Rain gardens optimize the value of any rain that does fall. The two techniques are not mutually exclusive—it is sound environmental practice to reduce or eliminate dependence on irrigation water in areas with regular water shortages, while at the same time introducing landscape design elements that will deal with periods of heavy rainfall that might normally give rise to flash flooding.

WATER AND SUSTAINABLE LANDSCAPES

Water brings our gardens and landscapes to life. It is multifaceted, with the potential to bring many different layers and meanings. The ideas and features in this book are relatively simple to create, and most contribute to making our landscapes, whether they be the home garden, the park, or the large-scale commercial landscape, more sustainable. Much of the rest of this book concentrates specifically on working with the landscape to reduce flooding and pollution problems, but water contributes in many other ways to making an environment that is good for people and for wildlife. Sustainable landscapes are often discussed purely in terms of environmental sustainability, but to be truly sustainable (i.e. lasting into the future without needing high inputs of resources and energy to maintain them) they must also be acceptable to the people who use them on a daily basis. In other words, bringing things down to the garden scale, your garden may be doing wonderful things for the environment, but if you regard it as ugly, scruffy and unsafe then it is not going to be sustainable in terms of its longevity, or the sense of meaning and pride that it gives to you. It is essential to always look upon ecologically-informed landscapes as being multi-functional and delivering multiple benefits, rather than becoming fixated on single issues—if one takes the narrow approach then so much wider potential is lost. Before we look in detail at the narrower concept of the rain garden we want to first consider how we can integrate water-conscious garden planning with a wider range of benefits that contribute to landscape sustainability, and we make no apology for placing people-centred benefits at the same level as environmental benefits.

Rain gardens are good for wildlife and biodiversity

Rain gardens promote planting—and that can only be a good thing. All the elements described in Section Two of this book only work because they are based on planting. Moreover, the more diverse or complex that planting, the greater the benefit. Simple mown turf or grass areas may be of some use, but generally speaking low monoculture vegetation will not be effective at either soaking up and trapping excess run-off, or in dealing with any pollutants or contaminants in the water. So, this is good news for plant lovers. Replacing paved surfaces, or intensively managed grass areas, with mixed naturalistic plantings not only results in overall reduced needs for maintenance, and inputs of fertilizer, water and energy, but will also greatly increase the wildlife and habitat value

Making water visible has additional side-effects: it provides an opportunity for lush naturalistic planting, and a haven for wildlife.

of a garden. Many advocates of rain gardens insist on native plants, but there is no functional reason why native plants are better than those from other regions of the world at doing the job that a rain garden is designed for, although of course there are strong moral and ecological reasons why native plants may be the first choice. All the rain garden features described in Section Two have enhanced habitat value compared to the simple plantings and vegetation structures that so often make up the garden landscape. Rain gardens are largely composed of flowering perennials and grasses, together with scattered shrubs—an ideal mix for encouraging a great diversity of wildlife.

It is easy to think of garden wildlife as being the visible and attractive things that we all like—birds and butterflies. But in reality garden biodiversity is largely about the things we can't see, or which are not prominent—insects and other invertebrates, hidden away beneath the vegetation or in the soil. The big things we like have to live off the small

Rain garden ecotone, Amstelveen, the Netherlands. Wetlands that exploit a diverse range of plants create the greatest potential opportunities for wildlife.

things that we cannot see. Rain gardens are particularly useful for supporting that greater biodiversity. Leaving the stems of perennials and grasses standing over the winter will provide a home for many invertebrates, as well as food for seed-eating birds. The diversity of flowers will provide nectar sources, particularly in the late summer and autumn.

The most effective wildlife-friendly landscapes take the form of 'mosaics' of different habitats: grasslands, wetlands, woodlands and scrub. Rain gardens provide an opportunity to work within this framework. Tall herbaceous and grassy vegetation in the rain garden areas may rise up out of a mown grass surround, and in turn the herbaceous plantings may interact with shrubby woodland edges. Such 'ecotone' structures maximize wildlife value—boundaries between two vegetation types or habitats being particularly valuable—enabling creatures from each habitat type to share restricted space. Where ecotones have a sunny south-facing aspect their value is at a peak—the warmth

Central Park, New York. This typical savannah landscape is a haven in a densely built city. People are instinctively and magnetically drawn to the water's edge. Stones and large boulders create good access to the water's edge and opportunities for informal seating.

promoting flowering and fruiting, and the basking of insects and other animals.

Ponds are the most straightforward wildlife habitats to create. Seemingly without any assistance they quickly colonize with a wide range of water animal life, and plants also will come in of their own accord. However, with some planning and design, very rich and satisfying water-based habitats can be created, often in the smallest of spaces.

Rain gardens provide visual and sensory pleasure

There is a theory that our fascination with water is with us as a result of our evolutionary history as a species, when in the distant past, humans developed from ape-like creatures that lived a semi-aquatic life around African lake and sea shores, existing on a very mixed diet of fish, hunted meat and fruits. We are left with an instinctive attraction to water in all its forms. Whatever the reason, if there is a lake, pond, stream, fountain or other water feature within a garden or park, you can count on that as being one of the things that people—and especially children—will automatically head for.

For many of us living in cities and towns, opportunities for contact with water are typically restricted to the park pond or fountain in the city square. In these urban environments, water has until recently been seen as a nuisance—to be controlled and contained. This change in attitude is somewhat ironic given that the presence of water was essential to the establishment of many settlements. Water was needed for drinking and cleaning, feeding livestock and transport. The rapid expansion of towns and cities in Europe during the 19th century changed this relationship. In the second half of the 19th century, "centralised drinking and waste water disposal systems" were introduced "in the wake of devastating choleras and typhus epidemics" (Kennedy 1997, p. 53) as cities rapidly became increasingly populated. Many cities have, perhaps without realizing, turned their back on this essential life force. In many industrial cities rivers, streams and brooks were frequently culverted to create more land for development. With the introduction of trains and new sanitation systems that piped water from outside the city, the river was no longer valued and often became a foul-smelling conduit for flushing away the city's filth. New development turned its back on this and the river was lost to the city. In recent years, however, with the regeneration of our inner cities, improving river water quality and the redevelopment of brownfield sites, water frontages are once again seen as a valuable asset that can enrich the experience of living in the city.

Rain gardens are good for play

One of the most rewarding aspects of designing with water is its huge potential to both animate and bring life to a landscape. In the USA, a study in a small New England town of children's attitudes towards their natural environment discovered that: "the most important qualities to the children were sand/dirt, small shallow ponds or brooks of water…" (Hart 1979). The rain garden provides the designer with new opportunities to explore how the gathering, transportation and storage and release of rainwater can not only achieve more environmentally sustainable designs but also create exciting and engaging play environments for children of all ages.

The potential of water to enrich the user's experience can best be illustrated by the proliferation of new and exciting water features in

Children are fascinated by water in all its forms.

Even crossing a boardwalk over wet ground is exciting. Bringing water into a landscape allows for interaction in a way that no other element promotes.

Photograph by Arie Koster

many urban regeneration schemes. In the UK there has, in the last decade, been a renaissance in the design and regeneration of the public realm in both pre-industrial and post-industrial cities. In many of these regeneration schemes water, typically in the form of fountains and jets, has been used as one of the principal design components to create the main focus and drama within the design.

In Sheffield, the local authority design team has used water as a binding element to link a series of new public squares and also as a means of connecting the city with its rich industrial heritage. In doing so, they have produced a scheme that has proved to be extremely popular with the public, especially with young children, who have come to view these new civic squares as a new water play amenity. The Peace Gardens, the jewel at the centre of the city's regeneration, best illustrates ways in which water might be used not only to enrich the cultural relevance of a design but also to create a range of different play opportunities, some of which had not been anticipated by the design team. Water is used in many different ways, including waterfalls,

Case study Peace Gardens, Sheffield, UK

The Peace Gardens were opened in 2000 and were the first stage of the city's regeneration master plan—the Heart of the City project. The design makes cultural references to the city's industrial heritage, which is based on the use of water to power the mills that developed along its rivers and streams. In order to control the movement, storage and release of water to efficiently power the mills, the rivers and valleys were modified. Weirs were constructed to raise the water level so that it could be abstracted via goits—these were typically stone-walled channels that directed the water into the millponds. The millponds worked like giant batteries, storing the energy of the water so that it could be released gradually in times of drought. These references can be seen in the cascades and rills that lead to the central fountain of the Peace Gardens. The play value of the rills was one element that had not been fully anticipated by the design team. One might expect that the narrow gently flowing channels, which visually connect the cascades with the fountain, might be used by children for improvised paper boat races. What was not anticipated was that in the surprisingly hot summer of 2000, families came to the Peace Gardens with their picnics, towels and bathing costumes. The children played in the fountains and lay flat out in the rills with just their faces poking above the water. The popularity of the gardens became so great that the local authority had to invest in additional security to ensure the safety of children crossing the road to the adjacent shops.

The Peace Gardens, Sheffield. Channels lined with ceramic tiles depicting water plants lead to the central fountains and make reference to the rivers and water power upon which the city was founded.

The Peace Gardens, Sheffield. Water flows like molten steel from the large copper cauldrons—a reference to the city's industrial past and present.

cascades, rills and fountains. Water is used to create a coherent design that links different elements, while also creating opportunities for public interaction and play, both intentional and—perhaps—unintentional.

This example helps to illustrate a number of points in relation to water and play. Firstly there is a tremendous appetite for opportunities in which people can physically interact with water rather than purely view it from a 'safe' distance. Sadly, research has shown that in cities, children have limited opportunities for informal play with water. In a study of three British urban areas, Robin Moore (1986) noted that: "considering the normal attraction of children to water the low rate of mention of 'aquatic features' in all three sites indicates a lack of water play …". Where there were once opportunities for outdoor water play in outdoor pools, paddling pools, rivers and ponds this has often been lost when local authorities have had to cut their maintenance costs or have restricted access for fear they might be liable in the event of injury. This fear of litigation is illustrated by the closure of the Diana, Princess of Wales Memorial Fountain in Hyde Park shortly after it opened, when three people injured themselves after slipping on the wet stones. The design, which incorporates a fast-flowing moat of shallow pools and rapids, was conceived as an interactive play feature. Sadly access is now restricted and the memorial is enclosed by a fence.

Secondly, Sheffield's Peace Gardens illustrate how we can create opportunities for play while also showing how water might be detailed in such a way that it contributes to the local identity and cultural significance of a design. This approach should encourage us to explore original and creative alternative ways of incorporating water while also enabling play to flourish in both anticipated and unanticipated ways.

What is exciting about rainwater harvesting is that it requires us to consider how we manage water when its supply can no longer be regulated by the turning of a tap, and where its disposal is not just a matter of pouring it down a drain. Rainwater harvesting has the potential for us to bring the water back to the surface and for it to animate the landscape in ways we cannot completely dictate but only anticipate and accommodate. There is therefore the potential to consider how we might create opportunities for play in a system that will require the rainwater to be gathered or harvested, transported, stored and finally released. Within

Photovoltaic cells in this 'solar sculpture' in the Mauerpark, Berlin, circulate rainwater, which is transported through a series of channels in the park into a water play feature. Even when dry the area is full of play potential, from the stepping stones to the pebbly beach.

This play area in the main street of a small German town takes the form of a ship. A natural stream has been diverted to flow around the ship, which sits on an island in the stream. Abundant planting makes for a high-quality experience.

this simple system there may also be opportunities for us to introduce human controls that allow children to determine when stored water is released to refill a pond, water a garden or perhaps flow into a sandpit or rill. We can also incorporate water pumps, which may be hand- or even solar- or wind-powered, to circulate water within the system. Indeed the addition of some form of pump adds an extra element that is normally only available with water play features powered from a mains supply. Without pumps, moving water can only be guaranteed during or after heavy rain (not the best play environment), or when released from a storage facility (where only a finite amount is available). Pumps enable circulation of the same water for a continuous supply.

Of course, there is a serious side to water play, which is why we are spending so much time concentrating on it here. By incorporating rain-water harvesting and enabling children to have some control over when and how stored water is released it is then possible for them to reconnect with the potential value of this finite and costly resource. While clean water is perceived to be always available we cannot expect our children to appreciate its true cost and its potential scarcity. However, we shouldn't just think that water play is for children—most adults find it equally fascinating.

Water and safety

Before going on to explore some examples of how water and play may be integrated within a scheme it is important to address some of the concerns over children's safety.

Firstly, what is the 'real' threat of water to the safety and well-being of our children? In 2002 the total number of deaths by drowning in the UK was 427 for all ages (Royal Society for the Prevention of Accidents 2002). Data on where these drownings occurred reveals that rivers and streams are the most dangerous places, accounting for nearly 40 per cent of all deaths. Garden ponds were the least dangerous location, accounting for 3 per cent of all deaths by drowning. Although the loss of any life is deeply regrettable, these figures begin to put in context the true nature of the threat, especially when compared to the number of road deaths in Great Britain for the same year, 3431, and which dramatically increases to nearly 40,000 when the figure also includes those seriously injured (Department for Transport 2005). Information gathered by the Department of Trade and Industry (DTI) on drowning in garden ponds between 1992 and 1999 reports that there have been on average eight deaths per year of children aged five and under and that of these deaths, 85 per cent were children aged one or two years. For the equivalent period, the number of road deaths for 0–4 and 5–7-year-olds

The Bijlmermeer is a large residential development built in the 1960s on the outskirts of Amsterdam, the Netherlands. The picture shows a large pool with beach and rope slide over the water. In today's increasingly litigious climate this approach to providing water play may no longer be acceptable. This design may be exciting but it is also vulnerable to potential vandalism. Broken glass or, worse still, hypodermic needles from drug users, may be discarded and partially hidden by the sand. Unless sand can be protected or rigorously cleaned each day it is no longer a safe element to be used in public recreation areas. This does not mean, however, that water cannot still be incorporated in inventive and stimulating ways.

An example from a public park in Düsseldorf, Germany, where bespoke play equipment has been installed to create opportunities for play and learning. The design provides opportunities to learn about how different civilizations tackled the issue of moving water from a lower to a higher level and also the physical effort that this requires. There is a water wheel, an Archimedes screw and a large ladle fixed on a pivot. The design is also very much about play. The water can be dammed with wooden blocks fixed by chains, to redirect the water along different channels.

The 2001 Potsdam garden festival in Germany included a children's play area incorporating conventional play equipment and water-based play, specifically designed for children of all ages. For older children there was a large rectangular pool with stepping stones, a floating platform and a jetty (see page 12). For younger children the design integrated sculpture, sand and water. Sculpted stainless steel fish heads protruded from a concrete retaining wall. A jet of water released from the mouth of the fish by pushing a button on the wall, next to the fish's head, poured into a giant sandpit. The wet sand created a different play opportunity, enabling the children to sculpt with the sand as if they were on a beach by the sea. The play area was incredibly popular and was thoughtfully designed for both children and adults. The adults were able to keep a watchful eye on their children from the café located at the heart of the play area. There was also a lot of seating throughout the play area, encouraging comfortable and casual surveillance and supervision. By thoughtfully designing for the needs of children and adults, the play area was popular and safe.

was 321 and 466, respectively. Information from the DTI goes on to reveal that boys are far more vulnerable than girls, accounting for nearly 80 per cent of all deaths for children aged 5 and under. The most interesting aspect of these statistics is that the majority of deaths did not occur in the garden of the deceased, which only accounted for 18 per cent of the total. The data would suggest that children are more vulnerable to drowning in a neighbour's garden (39 per cent), which they may have entered uninvited, or a relative's garden (29 per cent). By trying to protect our children from danger we may in fact be making them more vulnerable when they then encounter risks to which they have not been previously exposed.

Other northern European countries, including, for example, the Netherlands and Germany, appear to have a far more relaxed and permissive approach to incorporating and encouraging opportunities for water play within the public realm. However, there is clearly a potential risk to children when we incorporate ponds in our gardens. In the following section we consider how, through careful design, this risk can be minimized.

Safety and disease People may have understandable concerns about the use of open standing water in a home landscape and the possibility of encouraging pests and diseases. For example it is often assumed that rainwater collection facilities in a garden or landscape harbour just the right conditions for encouraging mosquitoes. However, apart from ponds, all the features described in this book are about temporary storage of water at the ground surface for hours or maybe days, and encourage infiltration of that stored water back into the ground. There is therefore insufficient time to allow breeding colonies of mosquitoes to develop.

A further concern relating to contact with harvested rainwater is the possibility of diseases or harmful substances building up which might, for example, infect children playing with that water. Here sensible precautions should be taken. Water that children may come into contact with should be from clean, non-polluted sources. Rainwater collected from a house roof in a rain barrel should be fine, so long as a tight-fitting lid is used, whereas direct use of run-off from a paved front yard used to park cars is not suitable. However, the use of planting and vegetation to filter and clean water can be very beneficial, so long as water can be circulated via a pump through planted areas. The example of the swimming ponds discussed later in the book, whereby harvested rainwater is used for swimming, and is kept clean through the use of wetland planting, is clear evidence of this. Therefore harvested rainwater that may be used for play features, or where children may have contact with it, should be circulated through wetland planted areas where possible. It goes without saying that so-called grey water and black water (domestic waste waters derived from washing or from toilets) should never be used where children are likely to be playing.

Rain gardens are good for the garden microclimate

Rain garden concepts can be effective in a number of ways at improving the garden or landscape microclimate. At the basic level, any approach that substitutes plants and vegetation for hard paved surfaces is going to cool the summer landscape. Paved surfaces store heat energy from the

Plants will provide a cooling effect in the harshest of surroundings.

Jets and fountains at the Generallife, Granada, Spain, humidify and cool the air in this hot climate, as well as making people feel cooler merely by their presence.

Case study Paley Park, New York, USA

Paley Park was completed in 1967 and was a pioneering example of a 'pocket park', taking up the space of the footprint of a single tower block amidst the high-rise development of the city. A 6-m (20-ft) 'water wall', a cascading vertical waterfall, dominates the space and creates a backdrop of sound that overrides the noise of passing traffic. But that isn't its only function—as the water falls it produces clouds of spray that cool the air as they evaporate and disperse. A light tree canopy provides dappled summer shade while allowing more sunlight in winter when the leaves have fallen. Walls covered with ivy again add to the cooling effect by reducing reflectance and radiation of heat from brick or stone walls.

View of the water cascade, which drowns out adjacent traffic noise, and (below) view into the park from the street.

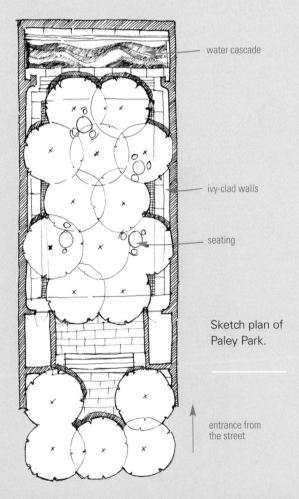

water cascade

ivy-clad walls

seating

Sketch plan of Paley Park.

entrance from the street

sun and re-radiate it at night, warming the adjacent air, and light-coloured surfaces reflect daytime heat radiation. Both processes are less effective with a plant layer. A tree or shrub canopy will provide shade, but will also cool the air through evaporation from leaf and stem surfaces—the heat energy required to change water from liquid to gas is removed from the surrounding atmosphere.

This cooling property has a practical application with the use of green roofs (discussed in detail in Section Two). As water evaporates from the roof surface and from the plants growing on the roof, heat is extracted from the roof surface itself, contributing to a cooling of the room area beneath the roof.

Finally, the simple presence and sound of moving water can be enough to make a person feel cool on a blistering hot day.

WATER IN A CHANGING CLIMATE

It is ironic that, with global warming becoming accepted fact, the impression is given that the major issue related to climate change is rising temperatures, and yet in reality the main effects of global warming will be played out worldwide through water. Shifting climatic patterns will not only extend areas where low rainfall and severe drought deny the basic ability of peoples to feed themselves, but also increase the frequency of severe and destructive storms and associated widespread flooding and contamination, while rising sea levels will steal the very land we inhabit. In short, water will be problematic both because we have too little of it, but also because we have too much of it. These opposites may well happen in the same place. In Western Europe over the past decade, for example, many countries have experienced exceptionally hot and dry summers, while in the same year being subjected to prolonged and sustained flooding events. In the past we have been shielded from major environmental disasters, often because they happen 'somewhere else', but also because technology has been able to protect the more affluent areas of the world from the more severe effects of shortages. It is likely that in the future this may no longer be the case, and moreover it is likely that we will, literally, feel the effects very close to home, in the way we live our lives, and the way that we use and manage our landscapes, whether they be our own small private gardens, or the wider public space.

It is very easy to sound pessimistic about our future prospects, but this is also an instance where the way in which landscapes are planned, designed and managed provides real and sustainable solutions to water-related problems. Moreover, this isn't just something for politicians, appointed public officials or for city planners, but an instance where small-scale actions at an individual level can have real impact. Residential areas, often with large proportions of garden space, represent significant coverage in a city, and collectively present important opportunities to influence how water works in our developed environments.

Water cycles

The way we use water in gardens and designed landscapes tends to treat it as an individual element—the dreaded term 'water feature' brings to mind isolated components, whereas in fact all water is part of a much bigger system. Every schoolchild is familiar with the idea of the water cycle—the endless sequence of events by which a molecule of water evaporates from the oceans, and condenses into clouds, which are then transported by atmospheric currents to deposit their loads over high ground as rain, hail or snow. Some of this fallen water then makes its way via rivers and streams to the coast and back into the ocean, while some evaporates back into the atmosphere (generally up to half the amount of the initial rainfall (Ferguson 2002)). The remainder infiltrates through the surface into the ground. When this water reaches an impermeable layer, such as clay, it accumulates in the saturated zone. Where this zone contains significant amounts of water, an underground aquifer is formed (Mueller et al. 2002). This large-scale combination of processes is usually treated as though devoid of human intervention and effect and as a system that is pictured as being in perfect balance.

In reality, water cycles operate on all scales and each site, be it an individual garden, a street, whole city or entire country, can be characterized by the way that water enters and leaves that area. Each area can be

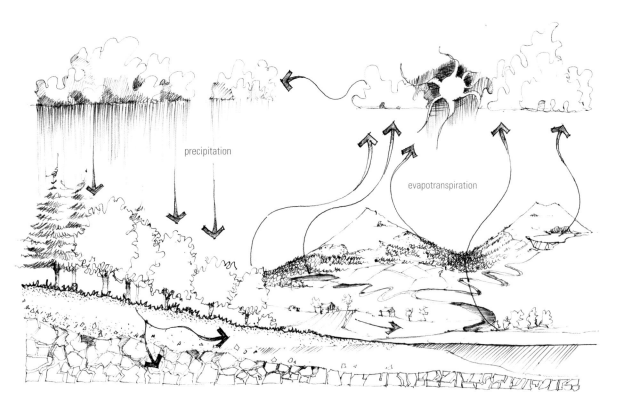

precipitation

evapotranspiration

viewed as a unit, with inputs of water from various sources, outputs of water via various routes, and various intervening things happening while in that area. When we compare how water behaves in a 'natural' area, such as a forest or a meadow, with how water behaves in a built-up area, such as a town or city centre, it is apparent that the activities of human development significantly change the patterns of water movement.

The main effect of built development is that the natural water cycle is short-circuited. Water that falls onto buildings or ground surfaces is shed rapidly into drains, which whisk that water away as quickly as possible into streams or rivers, or on to massive city water treatment centres. The natural processes of infiltration into groundwater, and evaporation back into the atmosphere, are reduced or eliminated and, as a result, there is often an unnaturally large amount of excess water following rain storms, causing the flooding problems with which we are becoming all too familiar. Several characteristics of the way we build and develop our houses, neighbourhoods and cities can be identified as contributing to this problem:

The classic water cycle. Evaporation of water from the sea, lakes, land and plants forms clouds, which then precipitate back onto the land. Some water will be captured in underground aquifers and plants or returned to lakes and oceans to begin the cycle again.

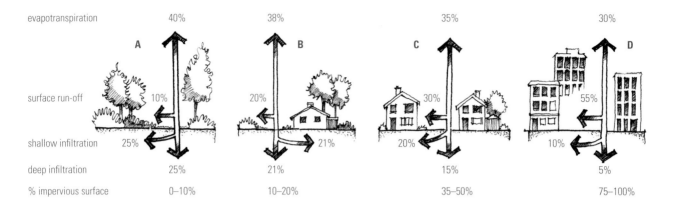

	A	B	C	D
evapotranspiration	40%	38%	35%	30%
surface run-off	10%	20%	30%	55%
shallow infiltration	25%	21%	20%	10%
deep infiltration	25%	21%	15%	5%
% impervious surface	0–10%	10–20%	35–50%	75–100%

Rainwater balance. With increasing development (A–D) and the associated increasing amount of impermeable surfaces there is a reduction in the amount of infiltration and increase in surface water run-off. The diagrams show the percentage of water leaving the landscape through different routes with differing amounts of impermeable surface.

(Adapted from FISRWG 1998)

- **The predominance of 'sealed' surfaces.** Sealed surfaces are those that form an impermeable layer over any underlying soil, preventing water moving downwards from where it falls as rain. Such surfaces are all around us: the roofs of our houses; the surfaces of our roads; the pavements and sidewalks on our streets; car parks; and the paths and patios in our gardens. One of the main effects of this preponderance of sealed surface is that water doesn't infiltrate where it falls, in the uniform way that rainfall over a field infiltrates, but instead it is shed rapidly from the main points of drainage from those sealed surfaces, leading to increased risk of flash flooding in heavy rain. The flow rate of water from development sites with high proportions of sealed surfaces can be two to three times that from predominantly vegetated surfaces (City of Chicago 2003). High rates of run-off may also lead to erosion and damage. Impervious surfaces not only move water away from the point at which it falls, concentrating it in other areas in large amounts, but they also prevent the natural process of infiltration into the soil, preventing natural processes of filtration and purification, and reducing the potential to replenish ground water.

- **Lack of vegetation.** Almost by definition, wherever sealed surfaces are found there is an absence of vegetation, because of that discontinuity between the ground surface and the underlying soil. Vegetation performs several important functions in the water cycle. It intercepts rainfall on leaf surfaces. Some of this intercepted

Land use	Coefficient of run-off
High-density housing	0.7–0.9
Medium-density housing	0.5–0.7
Low-density housing with large gardens	0.2–0.3
Sports grounds	0.1–0.3
Parks	0.0–0.1

The run-off characteristics of different urban land uses. The coefficient of run-off is a relative figure indicating comparative water shedding potential rather than absolute values.

(From Meiss 1979)

water may evaporate back into the atmosphere, but some of it will also fall onwards down to the ground. However, during heavy rain, the rate and amount of water reaching the ground will be reduced, thereby helping to reduce the risk of flash flooding. Vegetation also takes up water directly through its roots and some of this is also released through the process of transpiration (the continuous flow of water through plants from their roots to their leaves, and back into the atmosphere), while some water is also stored within the plant tissues themselves.

- Our drainage infrastructure—the network of drains, pipes and sewers—works very efficiently to remove excess rainwater far away from the point at which it falls, to prevent local flooding and unwanted ponding or puddling of water. Serious problems arise following severe rain storms when water shed from impervious surfaces is concentrated into the drainage system, which may not be able to cope with the surge of stormwater. In older drainage systems, the same pipes carry not only excess floodwater but also sewage and wastewater—these are known as combined sewers. Normally this poses no particular problem, but following storm surges the sheer amount of water can overload wastewater treatment plants, resulting in discharges of raw sewage into rivers and streams.

- Water running off sealed surfaces collects a wide range of pollutants, such as oil and other spillages from vehicles, animal faeces, sediment, dust, dirt, heavy metals, bacteria and other pollutants. Pollutant load is a particular concern where run-off drains directly into rivers, streams or ponds. Toxic elements can be harmful to aquatic life, but also increased nutrient loads (particularly phosphorus and nitrogen) can result in clogging algal growth, murky water and reduced availability of oxygen to other life.

In short, we have a problem with dealing with excess and contaminated water from heavy rain storms, but conversely we also have a problem with lack of on-site storage of water to help us deal with periods of shortfall. In periods of low rainfall, there may be very low flow rates in urban streams because there is little water left in the area and groundwater levels are low: cities may have local water shortages and aquatic ecosystems and habitats suffer (Ferguson 2002). We have an 'out of sight, out of mind' approach to dealing with these problems: for many Western societies water is seen as cheap and disposable. Clean water is mysteriously delivered to our houses and places of work along a network of pipes connected to some remote and beautiful place. Once used and contaminated it leaves, unseen, along an equally mysterious and convoluted journey to a place never visited.

Even at the small domestic scale we are not free of this sort of thinking. Areas where water accumulates in wet boggy patches have to be drained so that they can become usable, whereas in areas of low rainfall or water shortage we invest in irrigation to enable us to maintain lush growth, choice plants, or a fresh green lawn, and we artificially change the characteristics of our soils, incorporating plentiful organic matter to retain as much water as possible. What may seem to be small and isolated individual actions quickly multiply to create large-scale effects. For example, it is now well accepted that the continuous extraction of groundwater to irrigate lawns in Florida has seriously depleted reserves, and the leaching of fertilizers and pesticides in the run-off from those same lawns causes pollution of the local water courses. A recent trend of paving over British urban front gardens to provide off-road parking spaces has the side-effect of creating greater areas of impermeable surface that shed water into the street, contributing to urban flooding problems.

Maintaining a lush green landscape with excessive irrigation can be at considerable environmental cost.

At the level of the individual garden or yard, battling against nature becomes an expensive pursuit, and moreover, once embarked on this course, it is very difficult to leave it again, because our landscapes become ever more dependent on this continued support. However, at the municipal, city and regional scale, traditional engineering solutions to drainage and flooding problems are also increasingly costly and unaffordable, but the figures stack up to millions or hundreds of millions of pounds or dollars. The ongoing maintenance of these facilities becomes a continuous financial burden (Coffman 2002). It is also becoming apparent that these are not solutions at all. Building ever larger sewer pipes to take the extra run-off waters from built development; building ever higher flood defence walls to combat the increasing frequency of serious floods; containing rivers and streams in engineered banks and tunnels to contain their flows—all these concrete-based approaches at best postpone the onset of major problems, and at worst simply push the problem downstream onto someone else.

It doesn't have to be this way. In the last 20 years or so, a very different approach has come to prominence: an approach that attempts to restore the natural water cycle and celebrates the presence of water in the landscape, and that works with nature rather than against it. Known variously as 'low-impact design', 'sustainable urban drainage', or 'water-

37

Engineering the banks of streams and rivers to contain excess run-off is not only costly, it results in unnatural, over-controlled landscapes.

sensitive design and planning', these approaches are characterized by, wherever possible, making visible the movement and presence of water in the landscape.

Because the solutions in this book are dealing with the effects of impervious surfaces, namely roofs and pavements, we propose that plants are at the heart of the alternative way of working. The common feature of all the landscape and garden elements described in this section is that they involve soils, water and vegetation, often in close and intimate proximity to buildings, to not only achieve environmental benefit, but also great visual and ecological enhancement, and they provide the basis for a holistic approach to designing landscapes. Although the ideas have mainly been applied in larger-scale schemes, such as housing areas, highways and commercial and office developments, the water-sensitive philosophy offers an inspiring and meaningful basis to the design and management of individual gardens.

BIORETENTION

The basis of low-impact design is the concept of bioretention. Bioretention is a land-based practice that uses the chemical, biological and physical properties of plants, microbes and soils to control both the quality of water and the quantity of water within a landscape (Coffman and Winogradoff, 2002). A wide range of applications for bioretention have been developed that can be placed throughout a garden—these are dealt with in detail in the main body of this book. Although designed primarily for water management, making use of bioretention throughout a landscape brings with it all the advantages of a more environmentally-friendly design philosophy. Bringing plants, water and soil into built development has many other advantages:

- Environmental benefits, such as increased wildlife value, and reduced energy use and pollution, because atmospheric pollutants are captured in leaf canopies and the soil. The shading effect of plants creates a more pleasant microclimate.

- It **promotes** a sense of place and local distinctiveness, by responding to site topography and drainage, and by the use, where appropriate, of native plants.

- **The built surroundings** become more visually stimulating and dynamic.

- It **encourages** environmental stewardship and community pride.

- **Maintenance** requirements are reduced.

There is perhaps another benefit when we put water management at the forefront of our thinking—we are less likely to pave or tarmac over large areas of our land because it is more 'convenient' (Liptan 2002).

The Aztec Business Park, Bristol, UK, makes use of ponds and naturalistic planting as part of its landscape.

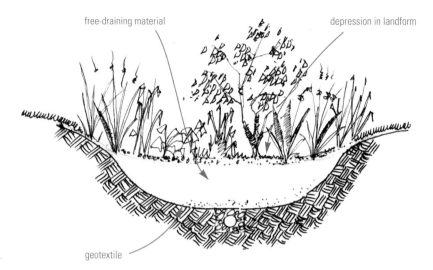

free-draining material

depression in landform

geotextile

General principles of a bioretention facility. These typically consist of shallow depressions of different shapes and dimensions that drain surface water run-off and encourage its infiltration into the soil. Where the natural soil is impervious, an engineered sandy or gritty soil may be substituted.

(Adapted from Coffman and Winogradoff 2002)

How does bioretention work?

Bioretention uses a simple model that provides opportunity for run-off infiltration, filtration, storage, and for uptake by vegetation (Coffman and Winogradoff 2002). All the features and facilities described later in this book rely upon the same mechanism for their action—excess rainwater is captured and filtered through soil, or 'substrate' if an engineered growing medium. Once the soil becomes saturated, water begins to pool on the surface, and is either able to infiltrate back down into the natural soil below and around the facility over time, or is drained away.

Control of stormwater quantity

The main purpose of the water-sensitive landscape is to reduce or eliminate the amount of excess run-off leaving the property or site—in so doing, pollutants held within the water are also contained within that landscape:

- Interception: the collection or capture of rainfall or run-off by plant leaves and stems, or soils, and the subsequent collection and pooling of that water in the bioretention feature.

- **Infiltration:** the downward movement of water through soil—this is one of the main functions of a bioretention feature.

- **Evaporation:** evaporation of water back into the atmosphere from plant and soil surfaces, and from pooled water. Bioretention features aim for shallow pooling of water to encourage maximum evaporation.

- **Transpiration:** the evaporation through leaves of water that is taken up by the plants growing in the feature. Generally, most of the water taken up by plants is transpired back to the atmosphere. The combination of these two processes from plant surfaces is known as evapotranspiration.

Control of stormwater quality

The ability of soils and plants to clean up contaminated water has long been recognized. Natural wetlands provide effective, free treatment for many types of water pollution. Contaminants may include organic material such as animal waste, or oil leakages. Inorganic material may include toxic heavy metals, or nutrients (e.g. from fertilizer use or from animal waste), which if it gets into streams or rivers can cause problems through over-promotion of plant and algal growth. The precise mechanisms by which this happens remain unclear and are likely to be a combination of several factors and processes:

- **Settling:** when water ponds within a bioretention feature, suspended solids and particles will settle out.

- **Filtration:** particles (e.g. dust particles, soil particles and other debris) are filtered from the run-off as it moves through soil and fibrous plant roots.

- **Assimilation:** nutrients are taken up by plants for use in their growth. Plants with high growth rates are particularly effective and can temporarily store mineral nutrients, until they are released when the plant dies back and decays. Plants may also take up heavy metal contaminants. By cutting back and removing the plant growth at the end of the year those contaminants are taken away. This plant-based cleaning up of contaminated soils and waters is known as 'phytoremediation'.

● **Adsorption:** the attraction of dissolved substances onto a surface—plant roots, soil particles and soil humus or organic matter—can 'lock up' contaminants by binding them to their surface.

● **Degradation and decomposition:** the breaking down of chemicals and organic matter by soil microorganisms. Wetland plant roots and stems increase the surface area for soil microorganism attachment. Moreover, wetland plants are adapted to growing in anaerobic conditions and can transport oxygen down through their tissues to the plant roots, which will be growing in waterlogged, oxygen-free conditions in saturated soil. The oxygen-rich conditions surrounding wetland plant roots again promote microorganism activity.

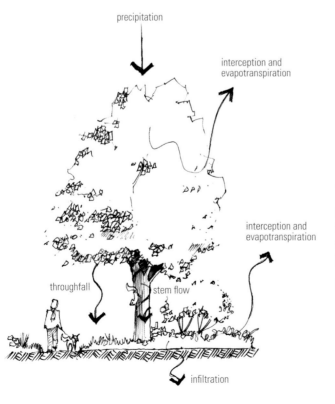

precipitation

interception and evapotranspiration

interception and evapotranspiration

throughfall   stem flow

infiltration

Plants and water management. The illustration shows different ways in which vegetation influences potential levels of run-off.

(Adapted from City of Portland 2004)

Although water purification is based on planted areas, it is not the planting itself that is the key component for water treatment, but the soil itself and the activity of microorganisms. However, plants have an important supporting function in increasing the permeability of the soil by creating secondary pore space; preventing compaction of the soil through their root and shoot growth; creating microhabitats for microorganisms on their root surfaces; and by channelling oxygen from the air down into the soil (Kennedy, 1997). Many nutrients are held in the wetland plants and are recycled through successive seasons of plant growth, death and decay.

The purification properties of wetlands have been utilized for the past 30 or 40 years in artificial constructed wetlands for the treatment of sewage and wastewaters, and this is sometimes applied in private gardens. We emphasize here that this book is not about treating sewage waters, or grey water from a house (the waste from non-toilet uses, such as showers, washing machines and basins), both of which have greater implications for human health than the processing in the landscape of rainwater run-off.

Grey water

Grey water is untreated household wastewater that has not come into contact with toilet waste. It includes used water from baths, showers, bathroom wash basins, and water from washing machines and laundry tubs. The use of wastewater from kitchen sinks and dishwashers is more problematic, and it does not include laundry water from soiled nappies. There is no theoretical reason why grey water cannot be used for irrigating home landscapes (but care should be taken with edible crops because of the danger of disease), and indeed in California it is a perfectly legal method of irrigation. In fact, the minute amounts of body oil, dead skin cells and soaps act as a mild fertilizer! Standardized grey water irrigation systems usually distribute the water from leaky pipes or containers that are buried underground. There is a potential danger from the ponding or collecting of grey water on the ground surface because of the remote danger of the development of pathogens and

This sustainable community in Stockholm, Sweden, treats grey and black water on site. Water from the houses is pumped through a series of wetlands and pools. Because of the potential hazards to human health, the wetlands are fenced off to restrict access.

disease-bearing agents. Because rain gardens work on the principle of ponding and infiltration this is a potential problem. We therefore advise that grey water can be passed through a rain garden sequence of elements as described in this book, but only infrequently, and only as an irrigation measure in very dry periods when ready infiltration is guaranteed. However, if grey water can be cleansed before it enters the rain garden sequence then that is a different matter. For example, a water treatment reed bed can take grey water direct from the house, and the outflow from the reed bed can then be fed into the rain garden sequence. Specialist advice should be obtained, and it should be borne in mind that such a treatment reed bed may need considerable space, and will not necessarily be an attractive feature immediately next to a house. But as a general point, we must point out that this book is restricted to the use of rainwater run-off, and the features described in Section Two are designed to work primarily with that run-off.

THE STORMWATER CHAIN

In essence, the aim of bioretention approaches is two-fold: to reduce the amount of impervious surface areas in order to reduce stormwater run-off, and to utilize the landscape and soils to naturally move, store and filter stormwater run-off before it leaves the development site.

Bioretention is applied in many and varied ways but the key is that, to be fully effective, an integrated approach is required that considers all aspects of the way that water comes into, moves through, and leaves any particular site or area. This integrated approach is often known as the 'stormwater chain', and comprises four major categories of technique (Coffman 2002):

1   Techniques that **prevent run-off from surfaces**.
2   **Retention techniques** that store run-off for infiltration or evaporation.
3   **Detention facilities** that temporarily store run-off and then release it at a measured rate.
4   **Conveyance techniques** that transport water from where it falls to where it is retained or detained.

This is therefore a spatial element to the use of bioretention facilities, and this is what makes the concept so exciting to us. The idea of a chain suggests linkage—one link in the chain joins onto another, and the more links are put together, the stronger the chain. But crucially, and rather obviously, a chain has a start and finish, and can even be joined end to end to make a circle or loop. The concept therefore opens itself up as an ideal basis for designing and managing a garden. The start of the chain will usually be a building—it can be the main house, or it can be a smaller garden building or shed. The end point can be at the lowest point in the garden, either a naturally-occurring low-lying area, or one that you make yourself. And the intervening links make up the bulk of the garden's planting and features—every single element within the garden can be tied into this concept. The chain doesn't have to be linear either—secondary chains and smaller chains can link in with the main chain in the same way that streams and tributaries join a main river.

In the USA, several pioneering cities and districts (such as Portland, Chicago, Seattle, and Prince George's County, Maryland) have developed guidance for developers and planners on how to incorporate stormwater features into the everyday landscape, replacing the standard widespread use of paving and hard surfaces with planted features, and substituting the ubiquitous commercial landscaping style of the same few common landscape shrubs that are used everywhere, with rich diverse herbaceous-dominated plantings, often based on native plant communities. This is working out to be cheaper than the engineer's usual fallback of more and more concrete-based solutions to stormwater-related problems!

Until now, this uplifting vision of how our surroundings can be planned to make beautiful, ever-changing landscapes that give back so much to the rest of nature has only been available to planners and developers, and then really only to a select few who can be persuaded to take on this approach.

It is our aim in this book to bring this new world of possibilities to the smaller scale of the home gardener, because it is in the garden that real experimentation and creativity can happen. Because the house is usually the starting point for the chain, we have a legitimate reason to say that this approach to working with a garden fulfils that cherished aim of fully integrating buildings with their landscape—the one can't work without another. If we start to design our buildings and landscapes in the way that we suggest in this book then we no longer need to look upon built development as the ecological evil that it is often regarded as—building can actually improve the ecology of a site compared to many other or previous uses!

Stormwater guides include a wide range of elements, features or 'facilities', all based on the simple bioretention model that we introduced earlier. They deal with the sequence and chain of elements that are incorporated into the landscape as one moves away from a building or a large area of impermeable surface, such as a car park or road. This is the same model that we will use in this book. The table and picture on the following pages show an ideal stormwater chain, and how it can be readily integrated into typical residential and public or commercial landscapes. We have indicated the relative position of each in relation

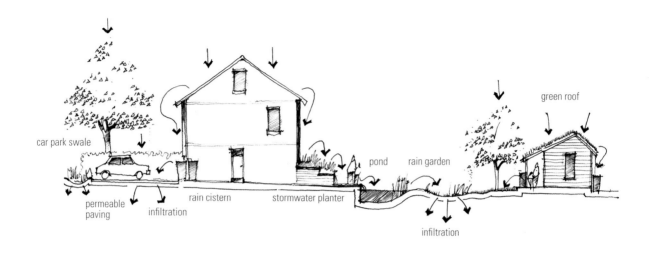

car park swale

permeable
paving   infiltration   rain cistern   stormwater planter   pond   rain garden   green roof

infiltration

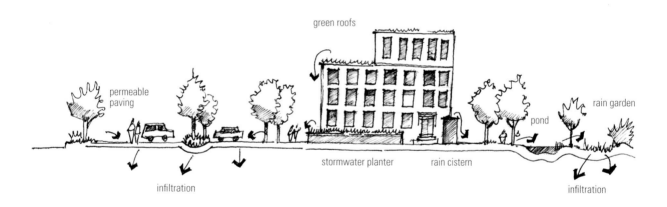

green roofs

permeable
paving   stormwater planter   rain cistern   pond   rain garden

infiltration   infiltration

to a building, and also summarized the main functions of each element, including the very important factor (maybe *the* most important for a home landscape) of 'amenity'—in other words how useful it is to have in your landscape in terms of visual beauty. Interestingly, all features, apart from rain barrels, score highly, and it could of course be argued that even rain barrels can be designed to be visually attractive. And again, nearly all of the elements have some habitat and biodiversity value. Each of these elements is looked at in detail in the next section of the book.

Residential and commercial stormwater chains. The two diagrams show how linked sequences of features can capture rainfall and release it back into the landscape at the residential (top) and commercial (bottom) scales.

	Immediate curtilage of the building			In proximity to buildings/built development			Some distance from buildings	
Component	Green roofs	Green facades/ vertical green	Rain barrels and water butts	Storm-water planters	Porous pavements	Rain gardens, infiltration strips	Landscape swales	Vegetation filters, constructed wetlands
Location/ application	Planted and vegetated roof surfaces	Climbers, green wall systems, vertical swamps	Rainwater collection directly from roof surfaces	Raised or at grade planters for rainwater storage and interception immediately at base of building	Hard surfaces that allow infiltration of rainwater	Planted rainwater collection areas	Parks, housing and commercial landscapes, urban infra-structure	Parks, housing and commercial landscapes, urban infra-structure
Run-off prevention	●		●	●	●	●		
Retention	●	●		●	●	●	●	●
Detention	●		●	●				●
Conveyance							●	
Filtration				●		●	●	●
Habitat	●	●		●	●	●	●	●
Amenity	●	●		●	●	●	●	●

Elements of a stormwater chain and how they can be integrated into typical residential and public or commercial landscapes.

It is easy to come across as being pious and sanctimonious when dealing with environmentally-friendly design, and also to be perceived as spreaders of doom and gloom, and dire messages of impending disaster if we don't all change our ways. It is also true to say that many examples of eco-gardens or sustainable landscapes come across as being very worthy, but also rather dull! So now it is time for confessions. Of course we sign up wholeheartedly to the need for much greater environmental awareness in all aspects of how we live our lives. But when it comes down to it, we have to admit that one of the main reasons for working in this way is that it offers so many opportunities to be interesting and creative in the way that we lay out, design and manage a landscape, and moreover, it gives real meaning to all the decisions that we make about how a landscape is planned—there's a story behind everything—it isn't just done because it looks good, or because it creates a satisfying pattern. And it does make us feel good to think that in addition we're doing something positive with that little bit of the earth's surface over which we have some control. We hope you'll feel the same way too by the time you get to the end of the book!

Above and opposite Water moves naturally in the landscape, collecting in lower areas and depressions. Rather than hiding or removing this natural drainage, why not exploit this natural tendency?

SECTION 2

The stormwater chain

In this section we look in more detail at the components of the stormwater chain, starting with those elements that can be used on or in conjunction with a building (be it the main house itself, or garden sheds, summerhouses or other smaller buildings), and then moving further into the wider garden or yard. It is possible to include all of the different elements that make up the stormwater chain in a single garden or landscape, but the size of area available will ultimately dictate what is possible. Using just one element will break the conventional drainage chain of roof or paved surface to sewer, and will therefore have an effect. Combining two or more elements together will multiply that potentially beneficial effect. So do not feel guilty if you cannot do everything—all elements are worthwhile. Many of the elements and components described here have been developed for large-scale applications and applied to public and commercial landscapes. All are applicable in home gardens, but may need to be modified to fit smaller spaces. Where we can we have included case studies and real examples to provide inspiration and precedent. Some of these examples are selected small details, while others illustrate what can be achieved when these principles are applied to an entire housing scheme or inner city regeneration project.

Each of the elements is introduced with some technical details and cross-sections to indicate how they are constructed and how they work. A series of case studies and design details then follows, before the next element is considered. Throughout the section, complete schemes and plans are introduced.

So, in the following pages we present a tool kit: a set of components that you can use and adapt. They can be put together in various combinations, and in ways that suit your particular garden or landscape. However, there are several principles that we suggest you bear in mind:

- Always think of the chain—that series of linked elements that build one upon the other and make sure that your elements are connected sensibly to what goes before and what comes afterwards—start from the fixed structures of buildings and work outwards in your scheme, and always work with the topography and levels that you are given, or which you are creating. If you don't feel confident initially in applying these techniques to the run-off from the house, why not adapt them to a smaller building such as a garden shed, greenhouse or garage? Some of the examples we have included specifically look at this scale of building.

- Make water and water processes visible, rather than hiding them way. It is much simpler to brush out the leaves that may block an open gully than it is to clear pipe buried beneath the ground.

- Above all, be creative and look for creative design opportunities everywhere! What materials do you have at your disposal and how can they be adapted to create a unique and original design? Reclamation or salvage yards can be an alternative starting point for interesting containers and materials. It also fits in well with the whole ethos of trying to work in a more sustainable way.

GENERAL DESIGN PRINCIPLES FOR BIORETENTION FACILITIES

In Section One we looked at the very general make-up of a typical bioretention feature or facility. Here we take things a stage further and look in a little more detail at the typical construction profile of an infiltration feature. Most of the elements described in this section modify this general structure.

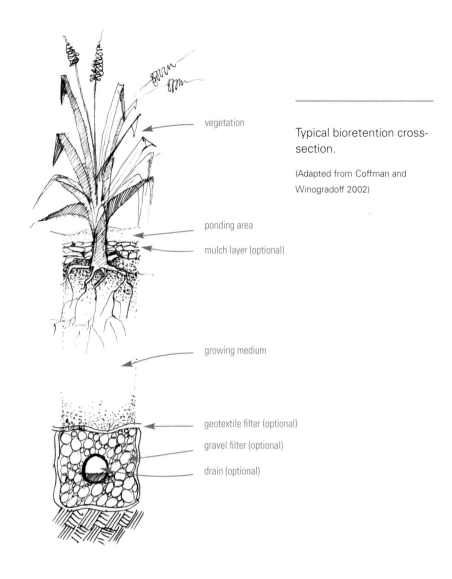

vegetation

ponding area

mulch layer (optional)

growing medium

geotextile filter (optional)

gravel filter (optional)

drain (optional)

Typical bioretention cross-section.

(Adapted from Coffman and Winogradoff 2002)

GREEN ROOFS

What are they? Green roofs are layers of living vegetation installed on top of buildings.

How do they manage water? Green roofs reduce the amount of water run-off from small to moderate storms, and also reduce the rate of run-off flow.

Green roofs are simply roofs that have had a layer of vegetation added to them. They are best known when used on a large scale: on schools,

Green roofs can be put on even the smallest structures: bird tables, kennels, recycling bins.

A garden shed with a green roof has become the focal point of this small garden.

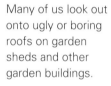

Many of us look out onto ugly or boring roofs on garden sheds and other garden buildings.

offices, factories and other big buildings. The opportunities at a small scale are equally great but have tended to be overshadowed by their larger counterparts. Garden sheds, porches, summerhouses, balconies, garages and small extensions offer great potential for planting green roofs. Indeed, for such a relatively modest feature, a green roof on a house or garden building can have an impact that far outweighs the area it occupies. If visible it immediately sends a signal about the intent behind a garden, flagging up environmental motivation in a way that other components simply fail to match. Moreover, creative use of green roofs can transform run-of-the-mill structures into the central focus of the garden. So often sheds are tucked away out of sight or, where they are prominent, can be extremely ugly, especially when viewed from above. Indeed, most of us who live in towns and cities look out onto our own, or someone else's, grey and monotonous shed roof. Greening up these surfaces not only improves the view, it also makes such utilitarian and functional buildings and structures into attractive focal points and features in their own right. Indeed, where garden or outdoor space is very small or restricted, a green roof might be the only chance of bringing plants and wildlife into otherwise hard or built surroundings. Let's turn these from embarrassments into the structures that we work the whole garden around! In a very small garden the summerhouse or shed becomes the focus of the entire scheme. Where space is more plentiful we can use green-roofed buildings to terminate views and frame vistas, or to create the theme for smaller enclosures or rooms. Making a green roof is a surprisingly easy thing to do.

In its widest sense, a green roof can include any sort of planting on a building, including traditional roof gardens (so-called intensive green roofs). However, most people now use the term mainly to refer to more lightweight, relatively thin roofs that are not necessarily intended for regular use or access, and which are the most 'ecological' in that they require minimal or no irrigation, and little in the way of maintenance. The most common types of green roof supplied by green roof companies are composed of sedum species, often supplied as pre-grown mats that are placed on top of the substrate layer or the drainage layer. These green roofs provide a reliable even cover and flower during the early summer months. There are many other options available, depending on the objectives of the roof. Green roofs provide ideal conditions for wild-flower meadows, particularly short-growing calcareous (chalk and limestone) types, and offer new opportunities for growing alpine plants more commonly used in rock gardens. It should be borne in mind that the term 'green roof' is in some ways misleading. Unless irrigation is supplied through summer it is unlikely that the roof will stay a pristine fresh green through a very dry season. For this reason the term 'living roof' or 'eco-roof' is often used to prevent misunderstanding.

As with so much else in this book, visual enhancement is an added bonus, the icing on the cake—green roofs also perform a wide range of environmental functions. Aside from the water management properties, which will be dealt with in detail shortly, green roofs provide a degree of winter insulation, they are effective at preventing the room below heating up on hot summer days, and are efficient sound insulators. The added habitat and wildlife value of greening previously barren structures is clear. Moreover, for those with a horticultural or ecological bent, garden green roofs provide rich opportunities for growing a very wide range of plants, both native and non-native.

Making a green roof

Green roofs are not a new idea. The turf or grass roofs of Scandinavian log cabins have been used for many centuries, if not millennia. It is worth looking at how these old roofs were made for clues about how similar roofs may be built on smaller structures. These roofs used local soils and vegetation (usually from the plot on which the building was built), and simple materials to construct the roof. Over a closely sealed wooden plank roof surface, layers of birch bark were laid to give additional waterproofing, and layers of birch twigs helped water drain from the base of the roof. Turfs were placed directly onto these layers, and the soil and vegetation layer held in place within a wooden plank or baton framework. These ideas, using modern-day lightweight materials, are adapted to achieve a feasible garden shed or outhouse green roof.

All green roof types consist of the same basic 'build-up' of a series of layers, and differ mainly in the depth of growing medium, and there-

Traditional Scandinavian log cabins use simple techniques to create green roofs, which provide insulation to the cabin in winter, and keep it cool in the summer. This example in the Skansen Folk Museum in Stockholm shows a typical old cottage. The roof is constructed with local materials such as birch bark, the plants and soil are held in place with wooden boards, and the roof plants are locally native.

fore the type of vegetation they support. Green roofs can be obtained and installed from commercial companies, or they can be designed and constructed on a more do-it-yourself basis, or a mixture of the two. We will first look at the standard construction of a typical commercial system and then consider how this can be modified. A typical commercial green roof consists of:

- **Waterproof layer.** The base layer of any green roof is the waterproofed layer of the roof. This must not only be waterproof, but also root-proof. Reputable green roof companies will provide a guarantee (usually 25 years) against leakage.

- **Drainage layer.** The drainage layer normally sits on top of the waterproof layer, and has the function of removing excess water from the roof. Most green roof plants are tough and drought-tolerant, and do not take to sitting in waterlogged soil. Commercial drainage layers take the form of preformed plastic cellular layers, and also incorporate purpose-made drainage outlets. Simple lightweight aggregates will perform the same function.

- **Filter mat.** A geotextile material usually sits between these two layers to prevent substrate from clogging up the drainage layer.

- **Growing medium or substrate.** The growing medium supports plant growth. Often referred to as substrate, this is usually an artificial 'soil' that is very lightweight. Typically, commercial substrates are composed of aggregate materials, such as recycled crushed bricks or tiles, light expanded clay granules, perlite or vermiculite, mixed with a small proportion (around 10–20 per cent volume) of organic matter such as green waste compost.

- The **vegetation** provides the living elements of the roof.

Section of a typical green roof build-up.

growing medium

filter mat

drainage layer

root barrier

waterproof layer

roof

Typical green roof elements are shown in this green roof under construction. A plastic cellular drainage layer is overlain with a geotextile filter mat, over which growing medium is spread.

Designing green roofs

Modifying a commercial system

You may wish to influence the appearance of a green roof by designing your own vegetation layer and/or using substrates of your choice. Most green roof companies will be happy to work with you on this, supplying the base layers and then leaving the rest to you. The simplest approach is to undertake the planting yourself onto a ready-supplied substrate, using either pot or plug material, a seed mix, or a combination of approaches. You may also source your own substrate. This is becoming common for the most ecological of green roof types, whereby locally derived waste or recycled materials are used (for example, the subsoil dug up when the building foundations are created, or demolition materials from a previous building).

Self-build green roofs

It is perfectly possible to construct a green roof yourself, using readily available materials. Of course, your plans must be completely watertight (literally!) if you are working for a client, for insurance purposes. If the roof is on a small, cheap structure such as a shed, or for personal use, then you can afford to be less stringent.

However they are constructed, there are two key considerations to bear in mind in green roof design: structural loading, and waterproofing. The most important factor for any green roof development is whether the building structure can take the additional weight that the green roof imposes. Where the green roof is being put onto a newly designed structure, then the structural support can be designed to take the required load, but where an existing structure is being 'retro-fitted' with a green roof, it is essential to obtain professional advice from an architect or structural engineer if in any doubt. Lightweight commercial systems have a loading of around 70–80 kg/m² (143–164 lb/ft²). However, where some diversity in the vegetation is required then loadings of around 100–110 kg/m² (205–225 lb/ft²) should be considered. This equates to a depth of around 70–100 mm (2.75–4 in) of growing medium.

It is also absolutely essential to ensure that the roof is completely waterproof before adding the green roof element. For complete peace of mind when working for a client it is strongly advised that you use a reputable roofing contractor. Even if you intend to use different substrates and planting than those supplied by a green roof company, it is strongly recommended that you employ the same contractor to install the waterproofing and also to place and spread the substrate onto the roof—in this way the guarantee will remain. Otherwise, you must clarify with the contractor how any subsequent work that you do above the waterproofing layer affects the waterproof guarantee.

Creating your own green roofs

The following notes apply to creating green roofs on small garden buildings that do not require adherence to building regulations or regulatory control. Such buildings include garden storage sheds, children's play houses, garden summerhouses, stables, pet and animal shelters, dog kennels, etc. The guidelines may also be followed on other surfaces so long as you are sure that the existing waterproofing and structural support is sound.

● **Waterproof and root-proof layer.** Over and above the existing shed or building surface it will be necessary to lay another waterproof layer. This gives added security and piece of mind, but most importantly will prevent plants rooting down below the green roof. Asphalt and bitumen roofs are very susceptible to damage from plant roots—if the surface is constantly wet then roots can break down the surface layer, but the greatest danger comes from plant roots getting in and exploiting the joins between the sheets of roofing felt. For a small shed roof a robust pond liner will suffice, preferably in a continuous sheet. If sheets must be joined together then it is imperative that the seals are completely watertight. For peace of mind you may also wish to cover the waterproof layer with a woven fabric such as roofing felt or a geotextile cloth. This layer will help protect the waterproof liner from any sharp objects in the soil or substrate. It also offers some protection from a careless plunge with the garden trowel while trying to remove a particularly stubborn weed.

Case study Extensive green roof, Sheffield, UK, with grasses and alpines (designed by Nigel Dunnett)

This green roof was created on a typical garden shed with a sloping roof, widely available from garden centres and DIY stores, and employs a method that requires no additional structural support. The aim was to transform a nondescript shed into the centrepiece of a small garden. After painting the shed, 7.5 × 7.5 cm (3 × 3 in) posts were set in concrete at each side of the shed and braced to it. A heavy-duty pond liner was laid over the existing roof surface to produce a root-proof layer. A butyl liner is preferable to some of the cheaper plastic liners. They are more flexible, less likely to puncture and do not degrade as quickly when exposed to the sun. A wooden framework was constructed that sits on top of the liner and rests against the posts, thereby taking some of the weight of the roof. The compartments in the framework are filled with substrate (in this instance a 50/50 mix of light expanded clay granules (LECA) and gritty John Innes Number 3 compost) to a depth of 10 cm (4 in), over a drainage layer of crushed aerated concrete breezeblocks. All construction materials are available from well-known DIY chains. Planting is a mix of readily available alpines such as thymes, dianthus and armeria, together with other seed-raised alpines and dry-tolerant short herbaceous plants and grasses.

Stages in the transformation of a typical small garden shed. The original shed was painted black, and wooden posts set alongside the longest sides.

In this case, drainage occurs though the base of the wooden framework. An alternative is to cut a hole through the roof and the liner where eaves extend beyond the wall of the shed. Care must be taken when cutting the liner to ensure that it does not tear. The liner and roof are then held firmly together with the aid of a plastic tank connector (available from any good plumber's merchant). A length of chain can then be dropped down through the hole and secured in place. Any excess water will then flow down the chain into a water butt or soakaway. It is advisable to cover the top of the hole with a small piece of geotextile (water-permeable material available from most garden centres). This will prevent the soil and substrate being washed down the drainage hole.

A plastic pond liner was placed over the roof, and a wooden framework rested against the posts.

The roof was planted with a range of readily available alpine plants.

- Drainage layer. This isn't necessary on a small sloping shed roof and it is perfectly possible to put soils or substrates directly onto the waterproof layer, so long as there is a slight slope to the roof to enable water to flow away. However, on sloping roofs it is important to consider some means of preventing slippage of the substrate and plants off the roof. One option is to construct a timber lattice or grid that can be placed onto the roof. The grid helps to prevent the soil from being washed down the slope. Once the plants have become established the roots will bind the soil and the grid may then be removed.

- Soils and substrates. The ideal substrate material for a green roof is lightweight and free-draining to prevent it becoming 'sour', but with some moisture-holding capacity so that it doesn't dry out too rapidly. There are many possibilities for the basis of a substrate: limestone chippings, crushed concrete, or brick rubble (from demolished buildings). Brick rubble is an ideal medium: it is a waste product, widely available, has good nutrient levels, good drainage and is attractively coloured.

Green roofs and water

Because roofs represent approximately 40–50 per cent of the impermeable surfaces in urban areas, green roofs have a potentially major role to play in reducing the amount of rainwater rushing off these surfaces. Green roofs influence roof water run-off in a number of ways. Water that falls on the roof can be absorbed into pore spaces in the substrate, or taken up by absorbent materials in the substrate. It can also be taken up by the plants and either stored in plant tissues or transpired back to the atmosphere. Some water may lodge on plant surfaces, and subsequently evaporate away. Water may also be stored and retained by the drainage system of the roof. By absorbing water and returning it to the atmosphere the roof reduces the amount of water available for run-off, and by storing it for a period before it runs off, it acts as a buffer

Run-off from a conventional flat roof and an extensive green roof over a 22-hour period.

(Redrawn and adapted from Kohler et al. 2001)

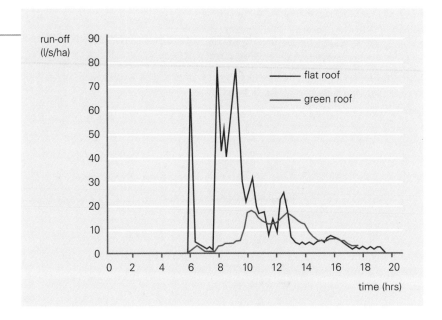

The green roof will absorb and gradually release rainfall—unlike a conventional roof, which rapidly sheds the water into the guttering and drains.

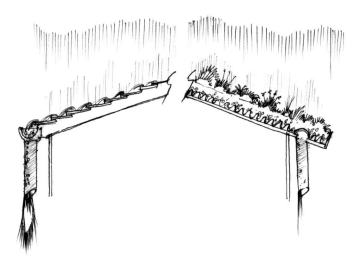

between the weather and drainage systems. Water stored by the green roof is gradually released over a period of time, so that the peaks of heavy rainfall characteristic of storms, especially summer storms, are evened out, and drainage systems are more able to cope with the amount of water entering the system. If there is a heavy downpour there may be too much water for the drainage network to cope with.

This can result in flooding and the possibility of sewage being forced through the system before there has been adequate time to treat it.

The graph on page 64 illustrates the typical effect of a green roof on rainfall run-off. The run-off from the flat roof mirrors closely the amount of rain falling on the roof and its intensity over the period of recording. The comparison of run-off from a typical conventional flat roof with an extensive green roof shows that not only is the total amount of run-off reduced (and the peak run-off reduced considerably), but that there is a delay in water draining from the roof, and the rate of run-off is relatively constant after the initial surge. This pattern is found consistently from experimental roofs across the world.

The storage capacity of a green roof varies with the season of the year, the depth of substrate, the number and type of layers used in its construction, the angle of slope of the roof, the physical properties of the growing media, the type of plants incorporated in the roof, the intensity of rainfall, and the local climate. It is therefore dangerous to generalize from the results of any particular studies, particularly if they were conducted in a different climate regime. However, most studies agree that yearly reductions in run-off of between 60 per cent and

Green roof stormwater retention over the period November 2003 to November 2004 in the Georgia Piedmont.

(Adapted from Carter and Rasmussen 2005)

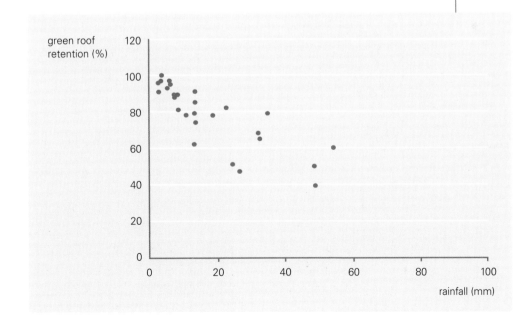

80 per cent are the norm. Moreover, peak run-off rates, as measured in mm per hour of run-off, can be considerably reduced. For example, the average rate of reduction on test green roofs over a typical April to September period was 51 per cent, with a greater rate of reduction for smaller storms (Moran et al. 2005). Further evidence for this variation in performance is shown by the graph on page 65, which shows how the run-off from a test green roof in Georgia varied with the amount of rainfall in a storm event over the period November 2003 to November 2004.

In general, the greater the rainfall, the less water the green roof retains. For small storms, the roof retains all or virtually all the rain that falls on it. This is mostly related to the moisture-holding capacity of the substrate—once the roof has attained its moisture-holding capacity, then any extra water will simply run off. There is a considerable difference between the amount of water retained in summer and in winter, due to the much greater amount of water that can be returned to the atmosphere in summer through evaporation and transpiration: retention rates in summer can be between 70 per cent and 100 per cent, but in winter may be only 40–50 per cent (Peck et al. 1999).

The role of plants and vegetation on the roof in reducing run-off is demonstrated in the table below. Even on the standard, ungreened roof there is some reduction in run-off when compared with the total amount of water that falls on the roof—this can be attributed to evaporation. Addition of a gravel layer results in some further reduction, but the greatest reduction is achieved with an added vegetation layer.

The influence of substrate depth and vegetation on the percentage of total rainfall running off the roof.

(From Mentens et al. 2003)

Roof type	Percentage of run-off from a green roof compared to that of an ungreened roof
Standard	81%
Standard with 5 cm of gravel	77%
Green roof with 5 cm of substrate	50%

Green roof types

Some plants grow naturally on old roofs and walls. Stonecrops, such as *Sedum acre* and *S. rupestre* for example, commonly find their way onto old roofs unaided in northern Europe, growing in very little or no soil, and rooting into cracks or joins between tiles. Similarly the houseleeks (*Sempervivum* spp.), as their name suggests, have traditionally been grown on slate and tile roofs, old walls and chimneys. Modern green roof technology simply allows these and similar plants to grow without endangering the structure and waterproofing ability of the roof below.

Depending on the depth of growing medium used, green roofs can support a wide variety of vegetation types (see table below). The shallowest, and the most common, types are known as 'extensive' green roofs, with depths of between 20 mm and 100 mm (they are known as extensive because maintenance is simple, low input and carried out across large areas). Plants suitable for extensive green roofs are tough, hardy and drought-tolerant, and tend to come from coastal, cliff, mountain and dry meadow habitats. The most common extensive green

Type	Substrate depth	Planting possibilities
Extensive green roofs	0–5 cm (0–2 in)	Simple sedum/moss communities.
	5–10 cm (2–4 in)	Short wildflower meadows. Low-growing drought-tolerant perennials, grasses and alpines, small bulbs.
Semi-extensive green roofs	10–20 cm (4–8 in)	Mixtures of low to medium perennials, grasses, bulbs and annuals from dry habitats. Wildflower meadows. Hardy sub-shrubs.
Intensive green roofs	20–50 cm (8–20 in)	Medium shrubs, edible plants, generalist perennials and grasses, turf grasses.
	50 cm + (20 in +)	Small deciduous trees and conifers, shrubs, perennials and turf grasses.

Summary of green roof types and planting possibilities.

(Adapted from Dunnett and Kingsbury 2003)

roof plants are those of the *Sedum* family and mosses. The extensive type of green roof is the most likely choice for garden use. On deeper substrates, 'intensive' green roofs can be used (intensive because maintenance input is much higher). These are the same as traditional roof gardens and require considerable structural support, and we will not consider them further here. However, there is an intermediate type, the 'semi-extensive' green roof. Semi-extensive green roofs combine some of the low-maintenance and low-input benefits of extensive roofs, especially if naturalistic or nature-like plantings are used, and can be more decorative, but do require greater substrate depths.

Planting options for domestic green roofs

Sedum roofs

Sedums are the most widely used green roof plants. They have many advantages in terms of hardiness and drought tolerance. Being succulent plants they actively store water in their tissues and shut themselves down during periods of severe drought to conserve water. Commonly used sedum species on green roofs include *Sedum album*, *S. hispanicum* and *S. reflexum*. All sedums are evergreen and most low-growing species flower for a relatively short period in mid-summer. White stonecrop, *Sedum album* (widely naturalized in the UK), and biting stonecrop, *Sedum acre* (a relatively common native of rock outcrops and old walls), have some of the most spectacular flowering displays. Sedum flowers are very attractive to bees, butterflies and other insects.

Sedum roofs can be established in three ways. Seeding is the cheapest option, but the seed is very small and it takes some time from germination to achieve a satisfactory vegetation cover. Sedums can also be established through strewing or planting cuttings or plugs—small sections of the plant will root easily. However, the most straightforward method for establishing cover on a small scale is through the use of a pre-grown sedum mat, which can be laid on top of the roof surface and whatever growing medium is being used, in much the same way as laying a carpet.

The garden pavilion at the Oase Garden, the Netherlands, where a community of artists live in a former monastery, supports a sedum green roof. Any excess water leaving the roof falls via a chain into a small rain garden. Any water running off this garden leads into a lake, which is connected to a series of smaller pools and a canal. The garden only uses recycled materials, and is managed to promote wildlife.

This park-and-ride scheme outside Ipswich, UK, has a sedum green roof on top of the passenger waiting area and ticket office. Excess water from the roof is fed into a water treatment wetland and settling pond.

Design: Landscape Design Associates

This green roof on a summerhouse shows sedums (mostly *Sedum album*) in full flower. A rain chain conveys excess water down to ground level.

Design: John Little, The Grass Roof Company. Photograph by John Little

Wildflower and habitat roofs

The conditions on a green roof (free-draining substrates with low fertility) create ideal conditions for the creation of highly diverse and species-rich grassland plant communities—rooftop wildflower meadows can be more successful than ones created at ground level. This is because frequent moisture and nutrient stress prevents the vigorous growth of aggressive and competitive plants that otherwise thrive in rich moist soils. In a small garden with restricted space, the rooftop may be the most appropriate place to establish a meadow. The most suitable types of wildflower meadow plant communities to adapt for green roof use are those that have developed over limestone and chalk bedrocks. The soils on which they develop are typically very thin—only around 10 cm (4 in) deep, and very dry because of the free-draining bedrocks, and support low-growing but very diverse plant communities. The parallels with the green roof context are clear.

Low-growing and creeping species such as cowslip (*Primula veris*), lady's bedstraw (*Galium verum*), bird's-foot trefoil (*Lotus corniculatus*), rock rose (*Helianthemum chamaecistus*), harebell (*Campanula rotundifolia*), hawkweeds (*Hieraceum* spp.), thyme (*Thymus drucei*), salad burnett (*Poterium sanguisorba*) and small scabious (*Scabiosa columbaria*) are likely to be successful. Depth of substrate will dictate what is possible— a minimum of 70–100 mm (2.75–4 in) is recommended. Green roof wildflower meadows can be established in a number of ways. Seeding is the most cost-effective method. Seed mixes should be sown at a rate of 1–2 g/m² (0.03–0.06 oz/yd²) for wildflower-only mixes, and up to 3–4 g/m² (0.09–0.12 oz/yd²) for grass and flower mixes. Mixing the seed with a quantity of sand before sowing allows the seed to be evenly spread across the surface. If sowing onto a sloping roof it can be beneficial to tack an open hessian or jute mat over the sown mix to protect the soil surface from heavy rain.

Wildflower plants can also be established using plugs (small plants in individual cells) or small pot-grown plants. This is more costly, but does give greater control over the planting composition of the roof. A combination of seeding and planting can be effective. Both seeding and planting are best done in the autumn and spring, and dry summer weather should be avoided. Where the visual effect of the meadow is important we suggest using a low proportion of grasses, or no grasses at all. Low-fertility substrates will give rise to short vegetation that won't require cutting back each year. Where growth is more productive or tall, a yearly cutting back and removal of cuttings will be required, to prevent matting or lodging of the died-back remains of the previous season's growth on the surface of the roof.

Biodiversity and habitat roofs

A particular type of habitat roof, often known as the 'biodiversity roof', has developed in Europe, which in many ways is the most ecological of all green roof types. It uses locally available or distinctive substrates (such as demolition rubble and crushed brick and concrete, or local subsoils, sands and gravels). Because they are locally derived they also happen to be cheap. Plants either come in on their own, or a locally appropriate wildflower meadow seed mix or plants can be used. This type of roof was originally developed in Switzerland by the researcher Dr Stephan Brenneisen, and is now receiving attention in

In the canton of Basel, Switzerland, flat buildings have to include a biodiversity roof, made with local stone and gravel substrates, and sown with a locally appropriate wildflower seed mix.

This small storage shed in a country park in London uses local waste brick rubble as the growing medium. The building is an educational feature in itself, with the walls infilled with logs, reeds and straw for invertebrate habitat, and recycled aluminium cans.

Design: John Little, The Grass Roof Company. Photograph by John Little

This summerhouse roof has a meadow character with grasses mixed with flowering plants. Chives, *Allium schoenoprasum*, are prominent in flower.

Design: John Little, The Grass Roof Company. Photograph by John Little

other countries. The concept developed as a means of conserving habitats that were being lost to development on the ground. For example, new building along the River Rhine in Switzerland threatens ground-nesting and wading bird habitat. By using gravelly substrates and soils taken from the River Rhine flood plain, similar habitats can be created on the rooftops.

In London, the same idea has led to 'brownfield' substrates, such as brick and concrete waste, being used on new buildings that are being developed on derelict or post-industrial sites. These sites are normally viewed as wasteland, but can in fact be very good for biodiversity, being undisturbed and based on free-draining and infertile materials. As a result, rare and protected species may be present. Recreating these conditions on the roof provides a viable compromise between development and conservation.

John Little of The Grass Roof Company, London, has installed around 20 green roofs and specializes in green roofs that promote plant and animal diversity. He has trialled many different substrates and now prefers a mix of brick rubble and fines—it is a waste product, and freely available. It has proved important for invertebrates and the nutrient level of brick rubble is relatively high, which allows a more diverse plant mix, and drainage is good. The Grass Roof Company design and build small projects including home offices, summer houses, bike shelters, storage sheds and verandahs. John says that the one thing he always tries to get across to promote living roofs is that it is simply a container and a bit of drainage to hold a substrate that will support plants—anything is better than tiles or felt. He does a lot of work with schools, in particular the local school now has one sedum roof verandah, one brick rubble bike shelter, one turf roof sports shed and one native/herbaceous storage shed.

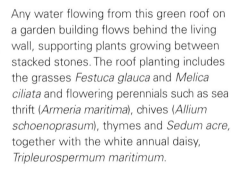

Any water flowing from this green roof on a garden building flows behind the living wall, supporting plants growing between stacked stones. The roof planting includes the grasses *Festuca glauca* and *Melica ciliata* and flowering perennials such as sea thrift (*Armeria maritima*), chives (*Allium schoenoprasum*), thymes and *Sedum acre,* together with the white annual daisy, *Tripleurospermum maritimum.*

Design: Nigel Dunnett

Decorative/ornamental roofs

On extensive substrate depths, a wide range of alpines and species from dry grassland habitats that might traditionally have found their place in a rock garden will grow very successfully on a roof—dry, free-draining conditions at the roots being a necessity. This is such a new field for horticulture that anyone can be a pioneer, trying out different genera or species for the first time. For example a wide range of dianthus have proved successful (the taller *D. carthusianorum* is a green roof favourite). The smaller *Erodiums* flower for months and months, well into early winter. Alongside the evergreen sedums, golden and variegated thymes provide winter interest, and the creeping crucifer *Hutchinsonia alpina* opens its white flowers in mild spells very early in the year. Another

Moorgate Crofts, Rotherham, UK. This roof terrace was designed for use by office workers and is an example of a semi-extensive green roof. The plantings had to look good throughout the year, require minimal maintenance and irrigation only in severe drought. Two different plant mixes were used: alpines and sedums in 10 cm (4 in) depth of substrate, and a naturalistic dry meadow of grasses and perennials in 20 cm (8 in) substrate depth. Different coloured stone mulches in swathes and bands provide visual interest in winter.

Design: Michaela Griffith (Rotherham Metropolitan Borough Council) and Nigel Dunnett

spring treat are the various pasque flowers (*Pulsatilla*), flaunting themselves with cowslips (*Primula veris*). Late colour is provided by exotic scarlet fuschia *Zauschneria* 'Glasnevin' and yellow *Penstemon* 'Mersea Yellow'. Bulbs are well worth a try, particularly the short species tulips from desert regions. At increased substrate depths (100–200 mm, 4–8 in), a very wide range of drought-tolerant perennials and grasses will thrive, for example a meadow-like mix of grasses (*Melica ciliata* and *Festuca glauca*) and a long-flowering succession of perennials of dry places—many scabious and scabious relatives can be successful. Annuals work well too—scentless mayweed (*Tripleurospermum maritimum*) and Californian poppy (*Eschscholtzia californica*) reliably turn up year after year.

CAPTURING WATER RUN-OFF: DISCONNECTING DOWNPIPES

With or without a green roof, there will be excess water flowing from the roof that needs to be managed and dealt with. In most instances, this water is a wasted resource, travelling straight from the roof through the downspout and into the drains and sewers. How can we capture this resource, which is delivered free of charge, and use it to not only enhance the sustainable profile of our garden but also enrich its aesthetic and experiential qualities? In this section we look at different methods of breaking the flow of water from roof to drain—in effect disconnecting the downpipes and putting the water to different use. There are several options for this temporarily liberated rainwater (City of Chicago 2003):

- Run-off can be sheeted across a lawn or other planting, such as filter strips.

- Run-off can be routed via swales and other landscape elements into detention or retention ponds, or rain gardens.

- Run-off can be stored temporarily in rain barrels or cisterns.

Clearly, downpipe disconnection must be considered carefully, and there must be sufficient outdoor space and planting to take the resulting run-off. We shall look first at the latter methods, aimed primarily at storing water.

Water storage

Until the advent of water treatment and distribution systems, people relied on rainwater and other natural sources for all their water needs. When people had to carry water from a spring, well or river into the house or out onto the garden there wasn't much room for spillage or wastage, and people developed inventive water collection and reuse techniques. We are now seeing a renewed interest in the centuries-old techniques, mainly because of the potential cost savings associated with rainwater collection, and the beneficial effects of using rainwater on plants compared to treated water (rainwater contains no chlorides, has zero hardness and contains fewer salts than municipal water).

Rain barrels and water butts

What are they? Medium-sized containers connected directly to downspouts.

How do they manage water? Rain barrels collect and store moderate amounts of water for small-scale non-drinkable uses.

Rain barrels and water butts have long been used to capture rainwater for use in the garden. Until fairly recently they tended to take the form of traditional wooden barrels, or metal or plastic barrel-shaped containers, and very often were fairly makeshift affairs that used any old leak-proof container, with the downpipes emptying directly into the barrel. Now, however, garden catalogues are full of a wider range of water butts that can make a positive aesthetic contribution to the garden. Particularly useful for smaller spaces are the flat butts that can be attached to the wall of the house or building. Instead of the down-

Water barrels can be ramshackle affairs, providing free water for watering plants in the garden.

Watering plants by can or bucket is more economical and efficient than indiscriminate watering by hose.

pipes emptying directly into the barrel, rain barrel diverters are readily available—these tap into the downspout but can be turned on or off when the barrel is full. Rain barrels should be fitted with a screen or tight-fitting lid to prevent breeding of mosquitoes.

The storage capacity of rain barrels can be quite large. For example, a 110 m² (1200 ft²) roof, using four 250-litre (55-gallon) rain barrels, one at each corner of the house, will store run-off equivalent to 1 cm (0.33 in) of rainfall over that roof—the equivalent of a moderate storm (City of Chicago 2003). Although not able to cope with severe storms, rain barrels can be an effective member of a longer sequence of stormwater management features. Their effectiveness is partly dependent on being regularly emptied for irrigation or other uses. But what happens when the barrels are full—for example during the winter when there is little need to use stored water in the garden? It is relatively common to join one barrel to another to produce a linked chain of barrels, the next filling up once the previous one overflows. But water barrels can also form the first part of the stormwater chain, making the first stop in that chain once water leaves the roof.

The water butt or barrel doesn't have to sit outside the building. In traditional greenhouses a water butt or large tank was frequently installed or built inside the building. Water from the roof is diverted from the downspout, through the wall of the greenhouse and into the

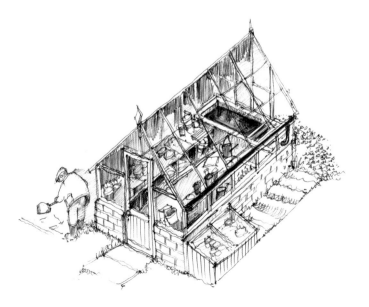

In this illustration we can see how a large tank has been built within the greenhouse and is filled with water from the roof. The water tank helps to moderate the temperature of the greenhouse during the growing season by releasing heat at night and increasing the humidity during the day. Any overspill from the tank could be directed to a particularly thirsty vine or soakaway. In winter the downpipe can be redirected to release the run-off into the garden.

water tank. This approach to rainwater harvesting has many advantages over using a mains water supply. It provides a source of water where it is most needed, without the expense of installing and maintaining a pipe from the mains. The stored water also helps to regulate the greenhouse climate by releasing heat during the night and maintaining humidity during the day. Finally the stored water is at the right temperature for young plants and seedlings, unlike mains water, which is typically too cold and may inhibit plant growth.

Disconnecting the downspout prevents water going straight into the conventional drainage system. In this student housing scheme in Portland, Oregon, disconnected downspouts empty into chambers that overspill through channels into planted infiltration basins. The channels are cleverly integrated into the paving so as not to interfere with walking.

In this example, water from the downspout is transported some distance from the building before being released.

Case study **Joachim-Ringelnatz-Siedlung, Marzahn, Berlin, Germany**

This 1990s housing development in Berlin is focused around a winding water channel, which acts as a link between a well feature at the start of the channel and leads, via a series of ponds, into underground water storage tanks. The stored water is used to irrigate the landscape. The channel is relatively shallow and is only full of water after heavy rain, but is a very attractive landscape feature in itself and is also used as a play feature by children. The base of the channel has been inlaid with the text of a poem, which is revealed as the channel winds its way through the landscape. This is a good illustration of design that has considered not only how the water gully will be perceived when it is flowing with water but also—and more importantly—what it will look like when dry.

Each house is provided with a galvanized metal rain barrel that collects the water from the roof—the water can be used by the homeowner. However, any excess water that cannot be stored in the barrel overflows into a granite channel that takes the water out of the garden and, via a series of basins and channels, into the central water channel. This design makes a very clear and visible link to all home-owners of the relationship between their own dwelling and the wider landscape in which their home sits.

The planting design within the public realm also reinforces the sustainable design principles that have played a significant part in informing the development of this project. Native plants, some of which include edible fruits and berries, have been used throughout the scheme to organize and structure space and improve habitat diversity. By incorporating edible plants in our urban environments we also help children to make the connection between food production, seasonality and consumption. In a world where our local supermarket supplies all things at all times, these simple connections with our environment are being lost.

In this sequence of pictures we see how each property has been installed with a simple galvanized water butt. The same detail has been used throughout the housing scheme, which adds to the coherence of the design and also helps to visually reinforce the sustainable ambition of the project. Overspill from the containers is transported to a large underground container via a series of open channels. This water is then used to irrigate the landscape.

**Case study NEC Gardeners' World 2000, Birmingham, UK
(designed and built by Andy Clayden and Nigel Dunnett)**

This demonstration garden was one of our first opportunities to experiment with some of our emerging ideas and interest in rainwater management, reclaimed materials and planting. We took our brief for the show garden as a small domestic backyard or garden that might belong to a family with young children. This very much echoed our own experiences so we were keen to explore a design that not only incorporated sustainable design principles but which could also be interacted with rather than only being experienced visually. The original sketch plan for the garden shows a small outdoor building, which might be a garden office or playhouse. The building is linked via a water channel and path to a seating area and garden pond.

We managed to construct the entire garden from reclaimed materials. The building was made from old floorboards and joists; the timber deck from reclaimed oak fencing rails that were painstakingly planed smooth; and the path and patio from local authority concrete pavers that we cut into slices to create a more interesting paving unit. The water channel was made from timber guttering or 'spout', to give it its local name (a distinctive feature on old terraced housing in Sheffield). It is durable and has the distinct advantage over plastic alternatives that you can lean a set of ladders against it without it cracking. Finding an appropriate water butt proved to be more of a problem. We rejected those available from DIY stores and garden centres. They were too large and would look out of place, and also they didn't fit in with the idea of using reclaimed materials. After some searching through reclamation yards we eventually found an old plastic water/header tank in a friend's back garden. This was a much better fit, it was the right colour—a pale sun-bleached blue—and more importantly it was also the right scale, tall and slim.

When we developed the design we wanted to show how the water would circulate through the garden by pumping the water from the pond back to the green roof and on through the system. We were keen not to use mains supply but instead created our own solar sculpture, which powered a small water pump. The solar panels were perhaps one

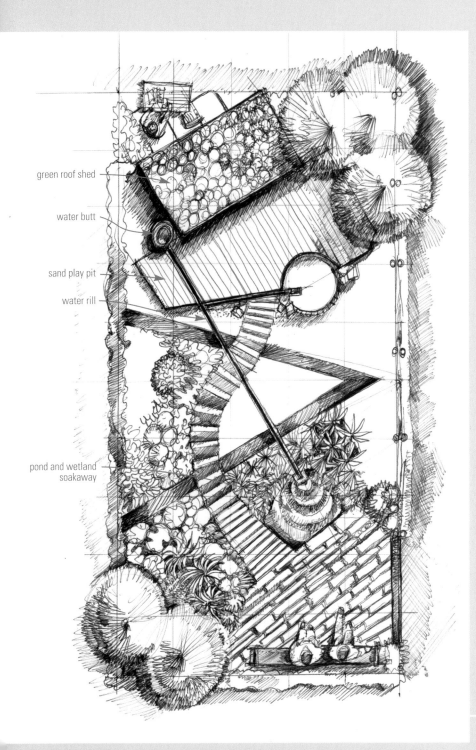

green roof shed

water butt

sand play pit

water rill

pond and wetland
soakaway

This image from
the completed
show garden
illustrates the
green roof,
water butt and
timber rill.

The plan shows a small
garden building, deck and
pond, which are linked by
water circulating through
the garden.

In this image we see the deep fascination that water has and its huge potential as a resource for informal play. The toddler has managed to dam the shallow channel while he plays with the water flowing from the tank.

Solar roof panels were adapted to create a moving solar sculpture that gently swayed in the breeze and powered the water pump, which circulated water through the garden.

of the most eye-catching elements of the design and generated a lot of public interest. They were mounted on flexible panels set within the planting, which gently swayed in the breeze.

Finally, one of the most rewarding aspects of the garden was its attraction for young children, who had no hesitation in stepping under the boundary rail and entering the garden. The integral sandpit was popular but the biggest crowd pleaser was the timber water channel and pond. As shown above, one of our own children managed to dam the channel with his nappy while he played with the water flowing from the water tank into the channel. This was important as it reminded us of the deep fascination children have for water, especially when it is flowing. Finding ways of capturing this quality by creating stimulating opportunities for play and interaction should be a guiding design principle. In the rain garden flowing water may only be a transient feature but it is this quality that can make it so exhilarating. It is reminiscent of the excitement that goes with a summer downpour, when the rain bounces off the pavement and for just a few moments the water surges down the roadside gullies in a torrent.

Rain chains

What are they? Linked chains that connect the roof gutter outflow with the ground.

How do they manage water? Transport and guide water to a desired location beneath the roof. Some run-off reduction through evaporation and spillage.

Rain chains encapsulate the ideas in this book—transforming functional items into aesthetic features. Rain chains ('kusari doi' in Japanese) have been used for hundreds of years to collect water from the roofs of homes in Japan, transporting it downward and finally depositing the rainwater into large barrels for household water usage. Japanese temples often incorporate large and ornate rain chains into their design. Rain chains guide rainwater visibly down chains or cups from the roof to the ground, transforming a plain gutter downspout into an eye-catching water feature. Rain chains hang from the hole where the downspout was, using the gutter attachment piece provided. When there is a mismatch because the hole is larger than the chain, a separate installation kit is used to reduce the hole and focus water downward onto (or into) the chain and also provides an outlet tube, preventing water from creeping along the underside of the gutter and dripping off.

Garm Beall is based in California and has set up RainChains.com to popularize rain chains in the USA. He recommends two types: link designs and cup designs. Link designs are the closest to the original form. "They tend to splash more than cup styles, and this may be important when they are considered for areas that are near doors, windows or walkways", says Garm. They are often used with modern architectural designs, but also look appropriate in rustic settings like cabins and log homes. Cup designs are an improvement over link chains in performance and efficiency. With open bottoms, they act as funnels, focusing the water from one cup down into the next one. Even in heavy rainfall, cup styles splash very little, so they can be placed anywhere.

The Japanese often put a ceramic or stoneware pot beneath the chain, which fills with water so that when it rains, the water drips from

Rain chains come in cup or link designs.

The rain chain makes the link between roof and ground. Although always visually effective, with heavy rain there is added interest as the water spills and splashes down the chain. It's not surprising that in Japan rain chains are used on temples.

Photographs by Garm Beal

the chain into the pot, creating an ever-moving display. Rain chains can be seen as a replacement for the downspout, transporting rain from roofs for use further into the landscape or garden.

Outflows and gullies (rills and channels)

What are they? The outflow is the point at which the water leaves the rain chain or downspout. The gully is a shallow channel set within the pavement or patio.

How do they manage water? Outflows slow the water down and capture it before redirecting it on its way. Gullies transport the water across a paved surface before releasing it into the stormwater chain.

The connection between the downspout or rain chain and the ground is an important one. At this point in the rain chain there is an opportunity to use the energy of the water for dramatic effect, to create noise, movement and drama. All manner of made and found objects can be used to personalize this detail, from a reclaimed stone sink filled with gravel to a crafted copper basin. The vessel may or may not accommodate the initial surge of water that rattles through the downpipe. It may overspill, only to be gathered by a shallow gully set within the paving.

Details may take their inspiration from other crafted objects. This drinking fountain detail (far left), set within the stone handrail of some steps, might be adapted to create a design for an outflow.

Westeveld Cemetery, near Velsen, the Netherlands. Rainwater is gathered from the roof, projected away from the building through a narrow metal pipe and allowed to fall freely into the pond below. The cascading water animates the space and surface of the pond and creates noise, which may disguise more intrusive sounds.

Photograph by Jan Woudstra

The outflow may retain a pool of water that helps to accentuate the slow drip, drip, drip as the rain gradually subsides. Alternatively the water may be released from the height of the guttering, falling noisily into a pond or large trough. Details might be reminiscent of the gargoyles found on medieval churches.

Around most buildings there is an area of hardstanding, which may include the pavement, a patio or parking area. For convenience, and safety in winter, these surfaces are typically designed to be well drained and free of water. In the rain garden water from the roof needs to cross this paved surface and cannot simply be directed below the ground into the traditional drainage system. It would be possible to pipe the water beneath this surface before returning it to the rain chain, although this would not be in the spirit of this approach to managing water and would also miss a huge opportunity to animate these areas of hard paving. By keeping the water on the surface it is simpler to maintain the drainage system. A gully can be swept clear of leaves, but an underground pipe may require the use of more specialist cleaning equipment. The gully can also be used to drain the pavement and patio as well as the roof.

Gullies are a
common method of
directing surface
water into the
drainage network.

A large pre-cast
concrete unit
can be used to
form more
organic
channels, which
can then be set
within a modular
paving system.

At the Welsh Botanic
Gardens in Carmarthen, a
narrow gully winds its way
down the central path.

Gullies can be formed in many different ways—they can be constructed on site or bought from a builder's merchant as a pre-cast unit. In most cities you will see areas of hard paving that include a standard gully detail, which directs the surface water to the drainage network (see page 89, top left). These systems can be used within the domestic context but they would be a missed opportunity to create something that was more personal and original.

Where a surface is constructed from modular paving such as concrete and clay blocks or even natural stone, it is simpler to keep the drainage gully running in straight lines as it reduces the amount of cutting required of the paving unit to accommodate the channel. Organic forms can be accommodated within a modular paving scheme where they have already been formed within a pre-cast unit (see page 89, bottom left). Where a more sinuous or organic gully is desirable it would be

Water is released from the rain barrel and redirected into a sandpit or allowed to flow into the garden to water plants or refill a pond.

better to construct this on site. For example the gully might be shaped from pebbles or small granite blocks, which can be arranged to create a gentle gully profile, set in concrete. The picture of the Welsh Botanic Gardens in Carmarthen on page 89 shows a sinuous gully flowing down a path, and also illustrates the potential of these details for play.

As already noted, the rain garden can introduce opportunities for play in many different forms. Gullies create an ideal opportunity for inventive play. They are safe and the water is moving. It is also possible to let the children control the release of water into the gully by including a water butt with a tap at the downpipe. Gullies may be temporarily dammed or the water redirected. When dry they might be used as a track for racing toy cars or playing with marbles. Gullies may also be used to connect other play elements with the patio. For example in the picture on page 90 the sandpit has been incorporated in the patio. Water can be released from the water butt and redirected to the sandpit.

Gullies may also serve in a more practical way. In the Moorish garden of the Mezquita, Córdobra (Spain), water is channelled through a series of gullies to each orange tree set within the cobbled patio. Water was a valuable and scarce resource that needed to be used wisely. The flow of water could be regulated by inserting timber boards into slots set within the rills. This same technique of regulating the movement of water can be seen on a much larger scale in the hillside villages of the Alpujarras. This approach to managing water could be adapted within our own gardens. Water could be directed to different areas of the garden. Alternatively, the gully could be modified to create a large shallow trough that could be temporarily blocked to retain the water. This might be an ideal solution to watering thirsty pot plants during the summer or while away on holiday.

Where it is not desirable to have a gully crossing a paved area it is still possible to keep the water open to the surface. The channel may be covered with a grate or the water can be allowed to filter through a series of narrow openings set within the pavement.

In the gardens of the Alcazar, Spain, water flows along a channel set within the paving, linking features and leading you through the landscape.

In the Moorish garden of the Mezquita in Spain, each tree set within the cobbled orange grove is individually supplied with water via a network of channels.

Stormwater run-off rushes along the gully beside the steps, rather than going into the drainage system.

Flow forms and a channel aerate the water while also creating an interesting sound and visual rhythm.

The design of the water outflows at The Cedar River Watershed Education Center, near Seattle, has taken its inspiration from the water catchment. The centre is a research facility that aims to promote environmental education. The water released from the building's green roof flows into a circular pool that contains rounded pebbles. This detail is reminiscent of the sculpted river bed where the force of the water combined with a large stone or pebbles erodes beautifully sculpted depressions in the bedrock. The rainfall is then gathered into a shallow gully that meanders across the open courtyard.

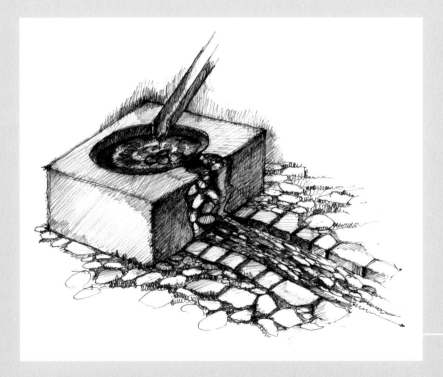

The design of the outflow from the Cedar River Watershed Education Center shows how a design may be both practical and original, taking its inspiration from the local environment while contributing to the building's educational objectives. This detail could be made by casting the block in concrete and then carving into the surface to create the overflow spout and expose the aggregate.

Disconnected downpipes feed water through gulleys into infiltration planters containing shrubby willows in a Berlin housing scheme.

Lower-lying areas in this scheme are cleverly used to take excess stormwater and temporarily fill with water, but for most of the time the lawns have a normal recreational function.

Stormwater planters

What are they? Above-ground planting containers that intercept water from the roof.

How do they manage water? Reduce run-off through infiltration, evaporation, transpiration and storage. Some pollutant removal.

Stormwater planters are one of the most exciting developments in rainwater management. Pioneered in Portland, Oregon, they are essentially above-ground boxes partially filled with soil in which plants are grown. The inspirational Portland Stormwater Management Manual describes them as 'structural landscaped reservoirs' (City of Portland 2004). In addition to reducing pollution, flow rates and volumes can be managed with these planters to moderate flows from buildings. Their great advantage is that they are sited directly against a building and can therefore be fitted into the smallest of schemes. Moreover, they can be integrated into a building design and contribute to the landscaping requirements of commercial, residential and industrial schemes—the structural walls can be incorporated into building foundation zones. They therefore provide a means of bringing planting and vegetation into the smallest of schemes, and also offer a creative alternative to the ubiquitous foundation planting at the base of American homes. On a larger scale, it is difficult to incorporate interesting and ecologically beneficial planting into commercial or residential schemes—so much planting is dull and monotonous and is comprised of the same small number of widely used shrubs or ground cover. Stormwater planters create an opportunity for the designer to improve both the visual and amenity value of a design because their environmental and engineering contribution to a project make them a necessary rather than desirable component of a development. Landscape architects frequently struggle to justify to a client significant expenditure on planting purely on its aesthetic contribution. Where planning authorities are actively encouraging these approaches to stormwater management there is a positive incentive for a developer to incorporate this technology. The designer is then able to justify working with a palette of plants that are typically absent from our arid urban environments.

Stormwater planters take rainwater straight from a building's roof because the downpipes empty directly into the planters. The first flush of water infiltrates through the soil in the planter. If the inflow rate exceeds the infiltration rate then ponding occurs up to the top of the wall level. This storage serves to attenuate flow. Excess water can overflow over the top of the planter into the next stage in the link of stormwater features, be diverted into the conventional drainage system, or simply infiltrate into the ground below the planter.

The Portland Manual recommends that the minimum depth of a planter must be at least 30 cm (12 in) unless a larger than required planter square footage is used. The minimum planter depth should be 45 cm (18 in). Special consideration should be given to the planter waterproofing if constructed adjacent to building structures. Depending on size, depth and context, planters can take large shrubs and small trees, but medium to small shrubs and ground cover are more common, with at least 50 per cent of the coverage to consist of grasses or grass-like plants. The design aim is that water does not remain in the planter for more than 12 hours—preferably the water drains in 2–6 hours. This is essential to prevent long-term waterlogging and anaerobic conditions, which are detrimental to plant growth. The plants chosen should be tolerant of periodic wet conditions, but are not water plants because the planters do not remain permanently wet.

Two types of planters can be constructed: infiltration planters enable water to infiltrate directly into the soil beneath, while flow-through planters overflow into the standard drainage system or the next stage in the drainage chain.

The infiltration planter intercepts water directly from the roof. Stones and pebbles dissipate the energy from the water, leaving the downspout to prevent gullying and soil erosion. The rainwater is temporarily stored in the soil and gravel at the base of the planter before being gradually released into the groundwater. If the quantity of water entering the planter should exceed infiltration, the excess water overflows the planter into the next stage in the link of stormwater features.

The flow-through planter creates a sealed container that gradually releases the water through evapotranspiration or into the next stage of the stormwater features. The flow-through planter may be fitted with a drainpipe that runs the length of the planter and diverts excess water either into the next stage of stormwater features or into the standard drainage network.

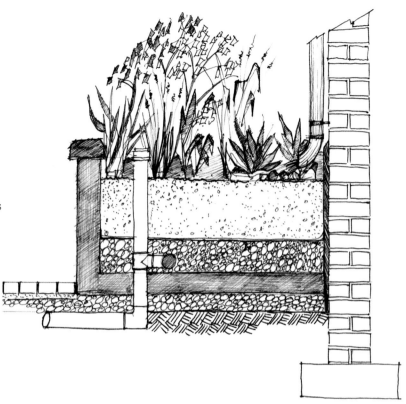

Stormwater planters create more than just an opportunity for rainwater management and planting. We should also think about ways in which we might use these structures to organize and arrange outdoor space. The sketch on page 99 shows a design for a garden patio. Rather than the stormwater planter running along the side of the building, it has been taken out into the garden to form two sides of the patio. This creates a more sheltered and intimate seating space surrounded by the vegetation. The sketch also shows how the overspill from the planter can then be directed into the pool. If damp-proofing the house is a concern the planter could be constructed a short distance away from the building. From our own experience of having young families, raised planters are also an effective method of creating some protection for plants from over-enthusiastic ball games.

The examples of stormwater planters we have shown tend to be quite formal and rectilinear. There is of course no reason why we shouldn't experiment with more organic and free-flowing forms.

Stormwater planters provide opportunities for rich planting around the base of a building and, combined with a green roof, significantly reduce or eliminate excess stormwater run-off from a site.

Photographs by Tom Liptan

Detail of the outlet into a stormwater planter. The disconnected downpipe feeds water directly into the planter.

Stormwater planters temporarily fill with water after heavy rain.

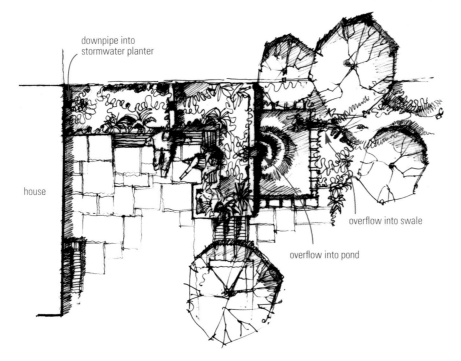

downpipe into
stormwater planter

house

overflow into swale

overflow into pond

Stormwater planters have been used to create a sheltered outdoor space next to the house. Bench seating has been attached to the walls of the planter. Water from the roof travels through the planters into a pond which overflows into a garden swale.

Rainwater harvesting

What is it? Large containers and cisterns (often underground) that receive run-off from roofs and other surfaces.

How does it manage water? Reduces total run-off through storage.

Rainwater harvesting represents the ultimate in capture and storage of rainwater for later use. This idea is an ancient one, evolving from arid regions throughout the world as a means of capturing rainfall for use throughout the dry season. Even a small roof has the potential to capture enormous amounts of water that would otherwise flow down the drain. This technology is becoming economically viable in many regions—for example in the UK, water companies are increasing water charges, regularly imposing bans on home irrigation because of water shortages, and increasingly charging for water use through water metering. While similar in principle to the idea of rain barrels, water harvesting is carried out on a larger scale, and the captured water can be put to a wider range of uses.

Water harvesting involves capturing rainwater from a house, barn or shed roof, filtering it and cleaning it, storing it for use in non-drinkable applications such as garden watering, toilet flushing, washing machine use and car washing. The water is passed through a filter that removes any organic matter, leaves, moss and bird droppings. Where roof surfaces are not available, parking areas or other sloping paved surfaces can be used, or even specially constructed surfaces covered with plastic or other impervious material.

Where domestic water use has been monitored using harvested water, it appears that enough water can be harvested in the British climate to account for all non-drinkable water needs. In a typical house this equates to a saving of around 50 per cent on mains water consumption. Costs range from £2000 to £3000 for a domestic system and cost around 5–10 p per week to run. With current water charges this gives a payback of around 3 years for larger-scale applications. Because water is stored in dark conditions there is no danger of Legionnaire's disease developing.

Storage tanks may be buried underground (often only feasible in new-build developments) or come free-standing. Free-standing tanks may be made of fibreglass or polyethylene. Both these are rather unattractive utilitarian structures, but can be painted, resurfaced or hidden behind wooden trellis. Much more attractive are galvanized metal 'farm tanks' that have a more traditional feel in many parts of the USA.

A centre for water harvesting initiatives is Austin, Texas. Home harvesting enthusiast Doug Pushard describes the reasons for his adoption of a rainwater harvesting system. "Although I use a drip irrigation system for my small city yard, the plants still consume lots of water during our long, sweltering summers." Austin receives about 81 cm (32 in) of rain a year, and on average 5–12 cm (2–5 in) of rain falls each month. "We don't have to worry about rainwater supply", says Doug, "the problem is that it all comes in buckets—all at once, and then nothing for weeks".

From his water bills, Doug estimated that his outside watering needs to be about 3785 litres (1000 gallons) per month on average. His rainwater capture system uses gutters that are connected by round downspout adaptors to underground 10-cm (4-in) diameter pipes that lead first to the filtration tank and then to the storage tanks. These are freestanding fibreglass tanks on a concrete pad, holding around 15,000 litres (4000 gallons). "This isn't enough to supply all our household needs each year, but it does serve a good proportion each month."

Doug Pushard and his home water harvesting set-up.

Photograph by Doug Pushard

Case study Folehaven residential community and Hedebygade Housing Block, Copenhagen, Denmark

The application of rainwater harvesting has been a key feature of a number of urban housing renewal projects in Denmark. At Folehaven, Copenhagen, a communal laundry, which serves a residential community of approaching 1000 dwellings, has been converted to a green laundry. A key feature of the green laundry is the use of rainwater, which is collected from the roofs and used to supply the washing machines, once it has been filtered.

Although this is a particularly large-scale application of this technology it has also been applied to a number of smaller regeneration projects. Located in one of the central districts of Copenhagen the Hedebygade residential block was selected as part of an urban renewal project that would focus on the application of sustainable design technologies. The renovated building and central courtyard, which was completed in 2002, incorporated a range of sustainable design features including solar heating, waste recycling and rainwater harvesting. Within the central courtyard, which can be used by all the residents of the housing block, a new community house was built. The community

The sketch plan shows the central courtyard surrounded by the housing blocks. The rainwater is gathered from the roofs and paved surfaces and transported via a gulley to a retention garden within the courtyard. The water is filtered and stored in an underground tank, then used to supply the washing machines in the communal building in the centre of the courtyard.

gulley around perimeter of courtyard gathers rainfall from roofs and paved surfaces

community building incorporating clothes washing facilities

rain garden temporarily stores water before releasing it to underground storage tank

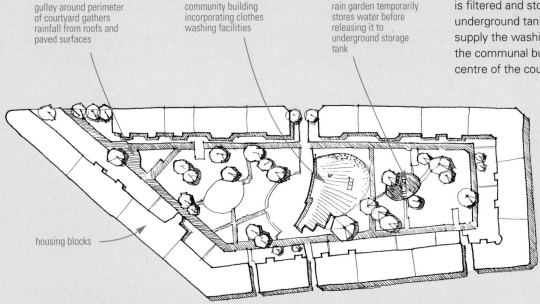

housing blocks

house included a community room, and kitchen and laundry, which used water gathered from the roofs and paved surfaces and transported via a network of gullies to a large underground storage tank. Before the water enters the storage tank it is gathered in a shallow open basin, lined with gravel, stones and plants. The basin has been designed to look like a pond although it is dry for much of the year. The introduction of communal resources is an effective way of not only implementing environmentally sympathetic water management techniques—it also has the advantage of removing large and costly items, like washing and drying machines, from a flat where space is limited.

Water is gathered from the roofs and paved surfaces into a gulley.

The communal building at Hedebygade. Rainwater supplies the communal washing facilities.

A rain garden temporarily stores the water before it is released to an underground tank.

Photographs by Anna Jorgensen

Infiltrating water

Porous or permeable paving
What is it? Paving materials and paving construction that promote absorption of rainwater and snow melt water.

How does it manage water? Reduces the quantity of surface run-off from small to moderate storms and some pollutant removal.

This book is primarily aimed at techniques and applications that permit greening of landscapes. Therefore we will not dwell overly on hard surfaces and materials. However, porous paved surfaces are an important tool for increasing the permeability of a landscape to water and they do in some instances improve opportunities for planting. Porous paved or reinforced surfaces are simply those that allow a proportion of the rain that falls on them to infiltrate through into the soil and groundwater, rather than all that water being shed directly into the drainage system. There are three main ways in which this can take place:

- Use of porous paving materials. Loose aggregate materials are the most obvious choice here, such as stone chips and gravel. Where the surface must take increased loading and must be solid rather than loose, modular paving blocks and grids can be used. These comprise structures with gaps and openings that can be filled with sand or soil. Often, such paving is sown with grass to produce a reinforced green surface. The plastic grid systems are particularly effective in lawned areas that are heavily trampled or require occasional vehicle access. The grid helps to protect the roots from being damaged and torn from the soil. They also help to prevent soil compaction, which impedes rainfall from infiltrating the soil. Absorptive asphalts and concrete are also available.

- Unsealed joints between paving units. One of the main ways by which water is shed from paved surfaces where there are no open gaps or joints between paving units. Such units are usually filled with a mortar mix. By leaving these open, or butting units close to each other to eliminate visible gaps, or by filling the gaps with a loose material such as sand, water can be encouraged to percolate through. Gaps set within the paving may also be less likely to clog than water-permeable paving systems, which may need to be pressure cleaned.

● Pervious bedding of paving. Rather than setting paving on solid and continuous sealed beds, such as mortar or concrete, paving can be set into permeable materials such as sand, hardcore (crushed stone or crushed reclaimed brick and concrete) and in some cases subsoil. For general garden applications where paths are not taking heavy wear, constructing paths to textbook methods of setting on to hardcore and mortaring in is not necessary. If heavy paving units are used they can usually be set directly on the soil surface or bedded in with sand. In this instance, paving will provide a cool moist root run for surrounding plants.

Permeable paving aids plant growth in two main ways. By enabling water percolation, plants are supplied direct with irrigation water—this can be very beneficial, for example for trees set in paved surfaces, as

Plastic grid has been used to reinforce the grass. The grid protects the roots of the grass and the topsoil from becoming compacted and impermeable.

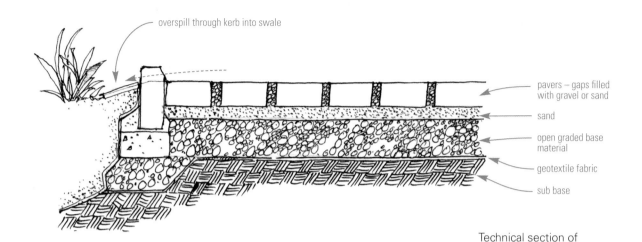

overspill through kerb into swale

pavers – gaps filled with gravel or sand

sand

open graded base material

geotextile fabric

sub base

Technical section of permeable paving.

shown in the image below. But more interestingly, unsealed joints between paving units allow plants to grow in these protected niches. Indeed, paving can be deliberately seeded with low-growing or creeping plants to fill those gaps. For any regularly used area of paving, those areas that receive the greatest traffic will remain plant-free because of the continuous disturbance, while those less frequently used will enable plants to grow. In this way a balance will be achieved between the amount of use and the amount of plant growth.

Trees in paving at the campus of Cork University in Ireland. The surface around this mature tree has been protected by a steel grid placed over gravel. This reduces the impact of soil compaction by pedestrians and promotes good water and oxygen exchange for the roots. Bollards have been placed around the edge to prevent vehicle access.

A 'green' car park in Berlin. The high density of tree planting means that the cars are all but invisible from outside. Porous paving enables plants to grow where the vehicle traffic is not so intense, and a relaxed maintenance regime allows wildflowers to colonize.

This modular concrete paving makes a durable surface, but the gaps can be filled with grass or other plants, and the units create a pleasing pattern in themselves.

Landscape swales

What are they? Vegetated channels and linear depressions.

How do they manage water? Swales temporarily store and move run-off water, reducing total run-off and flow rate from small to moderate storms. Some pollutant filtering capacity.

Swales are shallow, long, low depressions in the ground that are designed to collect and move stormwater run-off. As well as being a means of transporting water, one of their main functions is to allow water to infiltrate into the ground and to enable pollutants to settle and filter out. The aim is for them not to be permanently full of water, but to encourage accumulation of rainfall during storms and to hold it for a few hours or days while it infiltrates down into the soil, and/or is transported further to a detention pond or basin. Check dams installed periodically along their length promote greater infiltration and ponding up rather than rapid flow. Where swales are installed on sloping ground, check dams are useful in breaking the slope and preventing erosion caused by excess flow.

Swales are useful in storing and providing water for the immediate surrounding landscape and are one of the most effective solutions for

The water channel at Potsdamer Platz, Berlin, collects rainwater from the roofs of surrounding office blocks.

The stony base of this ornamental swale, planted with day lilies (*Hemerocallis*), promotes infiltration of water.

A plank bridge over a swale in a park in Stuttgart, Germany.

promoting conservation of water resources in gardens, commercial developments and along car parks, streets and highways. Diverse planting of shrubs, trees, perennials and wildflower meadows along their edges prevents evaporation of water from the swales, and the swale also provides irrigation for the plants.

The whole point of swales is that they are relatively narrow: wide depressions become basins and ponds. The Portland Manual recommends a water storage depth of around 15 cm (6 in), a maximum width of 60 cm (2 ft) for swales in private gardens or commercial developments,

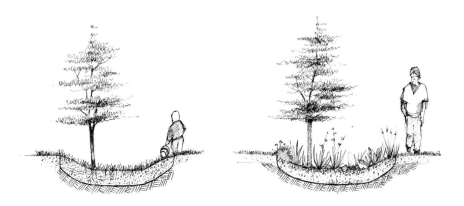

Grassed swale and vegetated swale. Water is transported along the swale and also allowed to permeate through the growing medium, which is separated from the subsoil with a geotextile fabric.

A grassy swale meandering through an ecologically-refurbished housing area in Augustenborg, Malmo, Sweden. Square concrete planters are intended to take larger plants.

Section of swale with drainage. If the swale becomes too full, excess water is directed into the drainage network.

and a maximum width of 1.2 m (4 ft) for public swales, such as those alongside highways. Where the natural soil type is relatively impervious, such as a heavy clay soil, gravel, grit or sharp sand can be incorporated into the top layer. There are two approaches to the planting of the shallow zone of the swale: *vegetated swales* and *grassy swales*. Vegetated swales have the primary function of promoting infiltration and can be richly planted: trees, shrubs and perennial plants. Naturalistic meadow-like communities are particularly appropriate. Grassy swales are better suited where water flow needs to be encouraged. Woody plants are

The overflow from this pond in a park in Västerås, Sweden, runs into an ornamental swale.

The swale is richly planted with lime-green *Alchemilla mollis*, upright *Ligularia stenophylla* 'The Rocket' and purple loosestrife (*Lythrum salicaria*).

A drain removes excess water from the swale in times of heavy rain.

usually, although not exclusively, kept out of the channel. In both types of swale the Portland Manual encourages the use of native grasses and wildflowers and recommends that swales should be designed not to require mowing. Where mowing is essential, it should not be done more than once a year—a natural-looking swale is no less effective than a manicured one (Liptan 2002). However, maintenance may be required to remove litter and debris in a public setting.

This ecological approach is in contrast to the approach taken in swale design in the UK, where invariably swales are composed of short mown grass, leading to an unnatural appearance and minimal habitat value (as well as a high maintenance requirement). Undoubtedly this lack of imagination is the result of a misunderstanding of the role of swales (a common assumption is that they are primarily water transport features), an emphasis on neat and tidy maintenance in public landscapes, and the fact that engineers are usually dominant in the design process. In tests conducted in Portland, swales vegetated with meadow-like mixtures of native grasses and flowering plants retained up to 41 per cent of the water flow through them, whereas identical swales vegetated with short turf grasses only retained 27 per cent of the flow (Liptan 2002). Several reasons may explain this result—the taller structure of the native vegetation may impede flow to a greater extent, and the native vegetation may have more robust root systems and also give rise to more organic matter, leading to greater moisture-holding capacity in the soil. Similarly, pollutant capture was better for the native vegetation, with 69 per cent removal of total suspended solids in the water for turf grass, and 81 per cent removal for native vegetation. If you are considering creating a grassy swale in your garden it may be worth checking that the profile is not so steep that your lawnmower blade hits the surface. If there is a concern that the swale may overflow, a drainage pipe connected to the main drainage network can be included.

Although swales are seen as impressions, they are of course part of a larger-scale landscape topography, and raised areas and slopes are necessary to shed water into the swale. Raised planting areas that are specifically designed to shed water into swales are known as *berms*.

Case study Prairie Crossing, Chicago, USA

Prairie Crossing is a 'conservation community' located 65 km (40 mls) northwest of Chicago, Illinois. Based on principles of energy-conscious building, creating a strong sense of place, and promoting commuting by train rather than car travel, the community strikes a balance between preserving the natural landscape and providing homes built in the local vernacular style. The development consists of 359 single-family homes and 36 condominiums. This is low-density development compared to the normal situation where thousands of houses are built on similar-sized parcels of land. Prairie Crossing includes an organic farm, school, community meeting house, and shopping centre.

The development covers around 280 ha (700 acres), of which 70 per cent is protected as open space—indeed the land was originally purchased to preserve environmentally-sensitive land from uncontrolled development. The open space is designed to provide stormwater man-

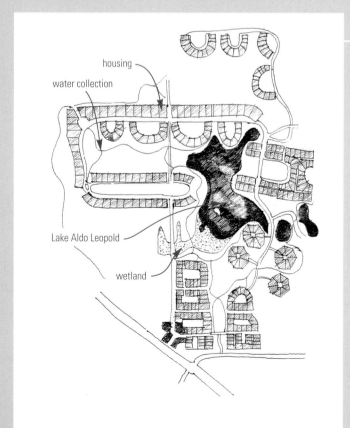

Site plan of Prairie Crossing, Chicago. Lake Aldo Leopold at the centre of the Prairie Crossing residential development gathers water from the surrounding areas.

agement and has been contoured to properly manage stormwater without the use of concrete culverts and other man-made stormwater discharge systems. At the centre of Prairie Crossing is the 9-ha (22-acre) Lake Aldo Leopold (named after the legendary conservationist) and a series of adjoining wetlands. A 'stormwater chain' sequence allows stormwater to drain slowly rather than being whisked away through a pipe. Stormwater run-off from residential areas outside the village centre is routed into swales planted with native prairie and wetland vegetation. These swales are the initial component of the treatment chain and convey run-off from roadways and residential lots into expansive prairies while providing for some infiltration and settling of solids.

Prairie Crossing,
Chicago.

Case study Berliner Strasse 88, Zehlendorf, Berlin, Germany

This scheme comprises 172 dwellings in housing blocks between two and six stories high. Two-thirds of the dwellings are rented social housing and a third private. The scheme was completed in 1993 and contains a day nursery, community meeting house and public green spaces. It is currently one of the best examples of creating a naturalistic setting for housing in a busy city environment.

Rainwater is harvested from the roofs and stored in underground cisterns after particles are filtered and it is UV treated—it can be used for non-drinking applications such as garden watering. Excess water drains into the collecting pond. Permeable paving is used where possible. A wind turbine and solar panels provide power to pump water from the pond to a central stream, which makes a water feature that all residents can enjoy. In a survey conducted in the late 1990s, residents said that their perception of the site was of a large car-free zone that seemed to be tailor-made for children.

At Berliner Strasse 88, water is pumped from the collecting pond to the head of the central stream and enters through a sculptural fountain feature.

Water flows through a rill into the main part of the site and on into the collection pond.

The stream/swale appears natural, but it is totally man-made. With or without water, it is a constant source of fascination for children playing in or near the water.

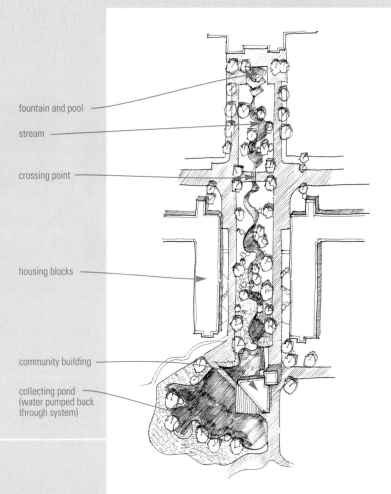

fountain and pool

stream

crossing point

housing blocks

community building

collecting pond (water pumped back through system)

Plan of Berliner Strasse.

Berliner Strasse.
Children play along
the edge of the
stream/swale.

The collecting pond
stores water gathered
from impermeable
surfaces.

Two specialized forms of swale have a good deal of application in and around the residential context: street swales and car park swales.

Street swales

One of the most exciting developments in stormwater management is that of street swales—small-scale features designed to take run-off water from the street. What makes them particularly interesting is that they are very similar to street narrowing constructions commonly used for traffic calming in European cities. Does this represent an opportunity to integrate planting and stormwater management with an accepted street structure and design to radically green streets, especially in residential neighbourhoods?

Water from the street is directed into these planted street swales in Portland, Oregon.

Car park swales

Most car parks and parking lots divide parking rows with raised kerbed islands, often containing trees. These can be replaced by a depressed sunken planted area within these large-scale paved landscapes. Water flows into these swales from the parking areas. Vegetation is essential to filter contaminants that may be in the run-off from the paved areas. Trenches lined with limestone chippings can also be used to trap any oil and petrol before the water enters the swale. Similarly, parking swales can be installed alongside domestic driveways and parking areas in front gardens and yards.

Car park swales provide opportunities for beautiful planting in these typically unattractive settings.

Photograph by Tom Liptan

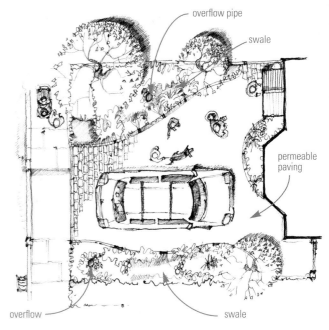

A vegetated swale runs alongside an area of parking. Some water may pass through the permeable paving or overflow into the swale. In small front yards where there may be limited opportunities to absorb all the rainfall, the swale may be fitted with a pipe to take excess water into the mains drain.

A sketch design for a small front yard where off-street parking has been included retrospectively. The design attempts to balance the need for parking and access while also retaining areas for planting and water infiltration. Trees have been retained because they have a positive influence on capturing rainwater.

overflow pipe

swale

permeable paving

overflow

swale

Filter strips

What are they? Gently sloping vegetated areas.

How do they manage water? Receive run-off from adjacent impervious surfaces, slowing the rate of flow, and trapping sediment and pollutants and reducing total run-off from small storms.

Filter strips make effective use of open spaces in a garden or landscape, providing they are not steeply sloping. Their main function is to absorb run-off from adjacent impervious surfaces, intercepting fast-running water and spreading it over a larger surface, thus breaking the flow. Filter strips can be used to take the run-off from disconnected down-pipes. Normal grass lawns can achieve this function very well, although mixed plantings will be more effective at promoting infiltration and pollutant capture. The wider the strip the better. Because filter strips operate on the basis of a sheet of water flowing over them, it will be necessary to prevent concentrated flows developing. So-called 'level spreaders' can be incorporated to evenly distribute the flow—these can take the form of a simple gravel-filled trench along the front edge of the strip. The main difference between filter strips and swales is that the filter strip is sloping and is designed ultimately to shed water rather than collect it. For this reason filter strips can lead into swales. In commercial settings where surface water may be polluted with heavy metals, for example motorway service station car parks and petrol stations, the filter strip may be filled with crushed limestone. The limestone helps to trap oil, petrol and heavy metals.

Water from the adjacent paved surface flows over the lawn and is evenly distributed across the grassed filter strip and woodland planting via a stone-filled trench or level spreader.

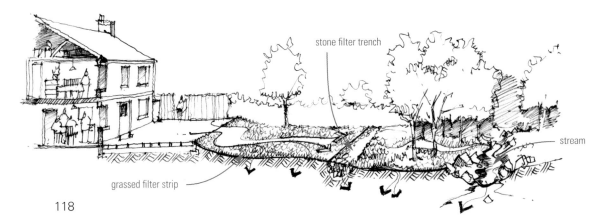

stone filter trench

stream

grassed filter strip

Case study Tanner Springs Park, Portland, Oregon, USA
(Design: Atelier Dreidetl/Greenworks PC)

Tanner Springs Park lies in the heart of a vibrant sector of Portland that has been regenerated from former industrial use. The park sits on a brownfield site, but the area was a wetland before its industrial development. One aim of the park is to re-interpret the history of the area from its wetland origins, and to reinstate water and wetland habitats as the main feature of the park. The park slopes from the highest level, which contains the typical park elements of mown grass, trees and planted beds, through large billows of stylized meadow, comprising tall masses of shimmering grasses, through wetland areas and marginal planting into a shallow water body. This gradient of planting reflects the gradient of moisture across the site, from wet to dry. Water is

Sketch plan of Tanner Springs Park.

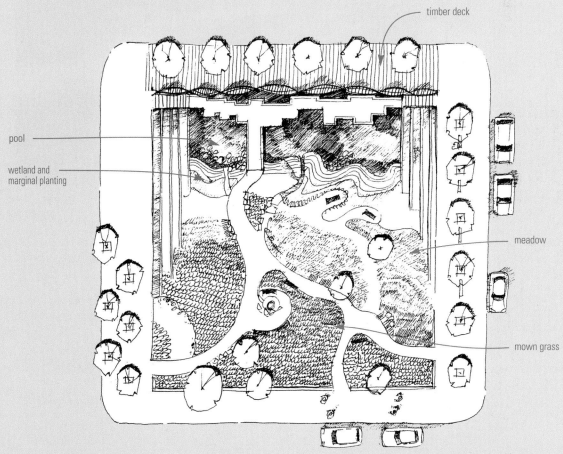

timber deck

pool

wetland and
marginal planting

meadow

mown grass

119

Water is gathered from the surrounding area and is filtered through a gradient of planting habitats before reaching a shallow pool.

pumped through channels and streams into the main water body. The calming, naturalistic quality of the wetlands and much of the planting is somehow sharpened and thrown into relief by some of the sculptural build elements of the park, such as the zigzag floating pontoon across the pool, and the waving sculpture wall formed from iron railway tracks, linking back to a former use of the site.

The park collects stormwater run-off from surrounding streets and pavements and filters through the sloping planted areas (filter strips) in which it is absorbed, but also cleansed, before finally collecting in the open water pool.

Retention ponds

What are they? Impermeable basins that will permanently retain water. Water may be lost from a retention pond if it overspills into a wetland or through evaporation.

How do they manage water? Ponds are one of the final elements in the stormwater chain, providing a final resting place for run-off water. An important function is pollutant removal.

Ponds are the most familiar of the elements in the stormwater chain, and the one that many residential landscapes may already contain. And yet even here there is a fundamental difference between a retention pond and a regular domestic pond in that, because they are designed to receive water run-off, any additional inflow into a pond will displace water that is already in the pond. This will mean that the water level of the pond will regularly go up and down. If the pond is already at full capacity, then any inflow will push water out of the pond and a regular overflow option is then necessary.

Retention ponds therefore mimic natural lake or wetland dynamics, in which water levels are not usually constant. And like these natural systems, ponds in gardens and designed landscapes can fulfil a very wide range of functions, from wildlife habitat through to aesthetic and environmental enhancement. This multifunctionality is enhanced through maximizing the planting opportunities that the pond creates. While the idea of regular changes in water may seem alarming in a domestic garden, this is an entirely natural phenomenon for a pond, and in fact makes for the greatest wildlife benefit.

Ponds will have a permanent pool of water within them, which rises following heavy rainfall and is then released over a period of time. Ponds can also have a pollutant removal function through settling, and by filtration by plants.

A form of pond that is more widely used in larger-scale or public landscapes is the 'dry detention pond' that dries out in periods of low rainfall, filling during rain storms. Because these are potentially less attractive than ponds with permanent water, attention should be paid to the visual quality of the banks: in any case they are not really suitable for a smaller-scale private garden, unless taking the form of smooth lawned basins that temporarily fill with water and can provide for multiple uses. A far more attractive solution is to plant the retention basins with planting that will tolerate periodic flooding or inundation. In effect one has then created a rain garden, and these are discussed in much greater detail later.

Small ponds have a very important role to play in protecting and promoting biodiversity. For example in the UK, ponds support at least two-thirds of all wetland plants and animals found in Britain (Williams et al. 1997). Creating new ponds and wetland features in the contemporary landscape mimics the natural process of the way that ponds form and pass through the range of successional processes to woodland; in other words ponds are seldom permanent and creating new ones continues an ongoing natural process. There is a problem, however, in creating ponds as part of the rain garden concept. The water for such ponds will usually come from two sources: direct from rainwater, and from surface water inflow (run-off). As we have already discussed, urban run-off is likely to be contaminated with high levels of soluble pollutants, heavy metals and organic compounds, all of which can flow directly into the pond. Not only do potentially toxic elements cause harm to wildlife, but increased nutrient levels in the water will lead to algal growth, green murky water and a generally unpleasant feature. For this reason it is essential that ponds come some way down the stormwater chain if taking run-off from sources that are likely to be contaminated and nutrient-rich, and should never directly receive drainage water from sources such as roads or heavily fertilized lawns—all the elements discussed previously have a water cleansing function and will act as buffer zones before water reaches the pond.

Designing ponds for safety

Because retention ponds are permanent water features they are potentially the most dangerous component of the water chain for young children and must therefore be designed with safety in mind. The following design guidance is intended to minimize any potential risks that they may pose:

- Restrict access to the water. This can be achieved in a number of ways. It is possible to design the garden so that the pond is physically but not visually separated from the rest of the garden by a low wall or fence. In this sketch, a low retaining wall contains the pond and wetland, which can only be accessed through a gate. This approach, however, is only a safe option if the gates are kept closed and bolted. A more secure option would be to permanently prevent a child from falling into the pond by installing a pond cover. These may detract from the aesthetic quality of the pond, although there are commercially available systems that can be installed just beneath the surface of the water and that can take the weight of a child or adult (see images on page 124).

- Design the pond with gentle profiles so that if a child does fall in they can easily walk or even crawl to the edge. Shallow profiles are also good for wildlife and establishing marginal plants.

- Where there is a deck or patio adjacent to a pond ensure that there is sufficient space for children to move comfortably around the edge of the pond and that all edging materials are securely in place.

The pond, which is supplied with water from the green roof, is separated from the garden by a wall and gated entrances.

A simple do-it-yourself framework placed over a small pond enables open water to be introduced into a small family garden. Even a small pond will support amphibians and insects and allow water plants to be grown.

Very gently sloping pond edges make the margins of the pond safer than a drop into deeper water. Here, permeable paving units give even greater assurance by providing a hard surface at the water's edge.

A plastic grid system secured to the edge of the pond can be positioned above or below the water level. The grid can support the weight of an adult.

Images supplied by Pond Safety Ltd

- Finally, the safest option is to ensure that children are supervised at all times, especially if they are visitors to the garden, who may be unaware of the potential dangers. To make supervision easier, locate the pond in an area of the garden that is visually accessible and near to where adults might naturally gather to have a coffee and chat.

Designing ponds for bioretention and for wildlife

Despite what has just been said, the functions of water retention, water cleansing and promotion of biodiversity are not mutually exclusive. One key point is to move away from the notion that a pond consists of open water and relatively steep banks, towards the notion of a pond being only partly about open water, but also about shallow (and very shallow) water, bare mud, and margins of vegetation. Indeed we can move still further and view ponds as part of a wider mosaic of wetland habitats that involve some amount of standing water—some of it permanent, some of it seasonal. The periodic rise and fall of the stormwater pond actually provides the opportunity for habitat creation.

In most natural ponds water rises and falls, creating a draw-down zone of variable wetness and high biological diversity. In standard pond designs this draw-down area is rarely considered and is usually restricted to a narrow strip at the water's edge. Extending the area of the draw-down zone considerably extends the habitat potential. A patchwork of hummocks and shallows maximizes diversity. Indeed very shallow areas of water are good for a wide range of invertebrates. Bare muddy areas are also valuable.

A mix of shallow water, deeper water, exposed mud and vegetation zones leads to maximum wildlife potential.

Case study Sutcliffe Park, London, UK

Sutcliffe Park is a district park (approx. 20 ha/49.4 acres), re-opened after redevelopment in 2004, which has facilities for the wider community such as an athletics track and kickabout area as well as extensive areas of new habitat creation. Previously the site was typical of many urban green spaces: a flat open space used mainly for football pitches, with little value to wildlife.

The redeveloped park is an example of how to significantly increase the biodiversity value of an urban park by linking creation and management of habitat and wildlife areas with wider environmental initiatives. The main reason for the redevelopment of the park was the need for a wider flood alleviation scheme in the area to protect the centre of

Sketch plan of
Sutcliffe Park.

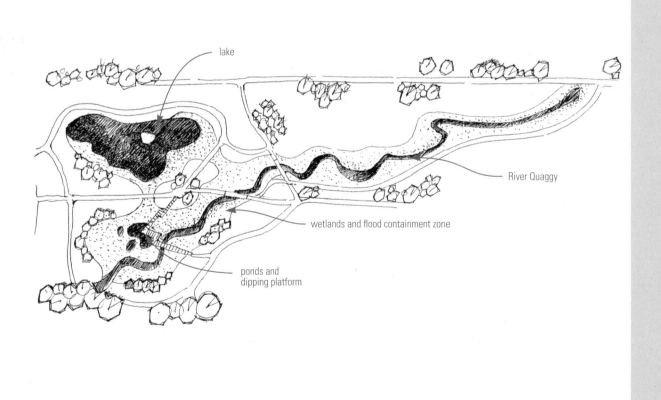

lake

River Quaggy

wetlands and flood containment zone

ponds and
dipping platform

Lewisham from severe flooding by the River Quaggy, as had occurred in previous years. A series of relatively unsuccessful engineering solutions had been attempted through the 1970s and 1980s and had completely covered and contained the river underground in a concrete culvert, leading to loss of natural habitats and reduced numbers of plants and animals, including fish, in the river. The re-development of the park centred around opening up the river again, and now, following removal of enough soil and subsoil to fill 35 Olympic-sized swimming pools, the river flows across the park in a sequence of meanders exactly matching those it had in the 19th century. At the same time, the surface of the park has been lowered and shaped to create an enhanced 'natural' flood plain where flood water will collect during severe storms. Instead of a flat and uniform stretch of metropolitan grassland, there is now a rolling landscape with a range of natural habitats to encourage as much wildlife as possible—the river itself, a lake, ponds (with dipping platform and boardwalks), wildflower meadows (both wet meadows and dry meadows at higher levels), an outdoor classroom, reed beds and a variety of native trees. At the same time, access for the public and for people with disabilities was increased. The scheme has great local support, largely because of the increased wildlife that it brings to the area.

A large increase has already been noticed in the number of people using the site, particularly families with children who like coming because of the water. Some of the gravelly areas by the river are very popular with children and some people think they have been made deliberately as large play areas, but in fact the children's enjoyment is a by-product of the habitat creation.

The reconfigured park contains large areas of open water as well as the 'daylighted' River Quaggy. The land has been re-profiled to accommodate different degrees of flooding, with areas that are wet all the time, and other areas that only temporarily hold water.

The scheme shows that new areas of open accessible water are perfectly possible in a public open space, and that perceptions of danger are not well founded.

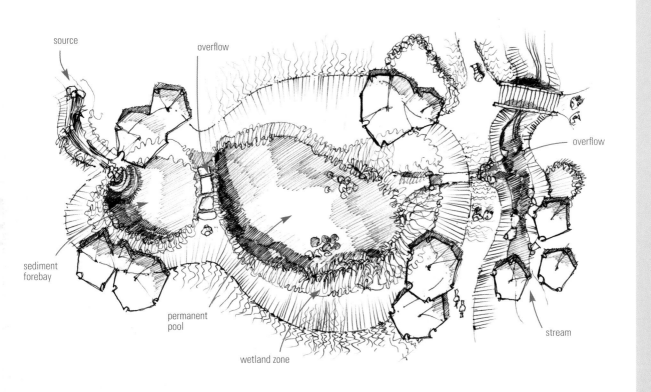

high-level water mark in autumn and winter

low-level water mark in summer

A retention pond may include a forebay area to encourage sedimentation and an outflow for excess water.

Comparison of the draw-down zone of traditional ponds and natural ponds. The zone doesn't slope evenly but is composed of a mix of higher and lower areas, giving rise to a complex mosaic of conditions.

(Williams et al. 1997)

source

overflow

overflow

sediment forebay

permanent pool

wetland zone

stream

The main point to bear in mind is to minimize the extent of dramatic changes in level as stormwater surges flood into the pond, and then drain away. Ensuring that a good half of the pond permanently retains water will provide sufficient stability for most wetland wildlife. As noted above, increasing the potential surface area by the use of shallows and the draw-down zone will also enable additional influxes to be widely distributed. And, again as noted above, by preventing direct inflow from the main sources of stormwater (i.e. by not disconnecting downspouts direct into a pond), the speed of water entering the system can be significantly reduced—surging water will disturb the pond and its sediments, leading to clouding and algal growth. So, again, the pond should be placed further down the chain if it is to receive stormwater run-off.

Ponds in a garden setting will rarely be designed for wildlife alone: people will want to see the water and gain access. Where wildlife is an important factor it is common practice to make part of the pond edge easily accessible to people, and to retain part of the edge, if possible, as a limited access point, to provide undisturbed or less disturbed areas of cover—for example to encourage bird nesting.

A path on the far side of this pond allows for fishing access, while the opposite bank has plenty of cover for nesting birds.

Ponds designed for wildlife where there are also lots of people around often have accessible edges or banks to allow views across the pond, or for educational activities, and a non-accessible wildlife zone. Here a pond-dipping platform sits opposite a non-intervention wildlife bank.

Decking viewing points result in minimal disturbance to the edge of the pond.

Boardwalks are exciting and can be a strong visual element of a wetland area, providing year-round access to marshy, boggy or wet areas.

Where it is known that surface water run-off inputs into the pond are likely to be nutrient-rich or polluted, all is not lost. The shallow margins and draw-down zone can be focused on. In polluted waters, submerged aquatic plants rarely thrive (Williams et al. 1997), and the water has the main function of settling out suspended solids. In the shallows there is a much better chance of marginal and emergent plants surviving in most water qualities.

Log piles make a structural feature within this wetland theme garden at the London Wetland Centre. Dead wood is a valuable habitat for a wide range of invertebrates.

The Western Harbour development in Malmo was the site of an international housing exhibition in 2001. The master plan for the site was based to a large extent on the management and use of rainwater run-off. Housing areas were developed around courtyards. The water that runs off roofs and pavements is channelled into strongly designed pools in each of these courtyards, and each pool is richly planted. Excess run-off is also sent into a large lake that is the focal point of a new park. The park features large areas of reed bed-like planting, and areas of wet woodland. Developers were invited to submit proposals for different parts of the site within the master plan. To get permission to build, each developer had to gain a maximum of 150 'biodiversity points' for their development, made up from a number of possible features that they could include, each of which gained 10 or 15 points. Potential features include:

- Climbing plants on walls
- Green roofs
- Bird boxes
- Native planting
- 1 m² (10 ft²) pond for every 5 m² (50 ft²) of sealed surface
- Amphibian and insect habitats
- Bat and swallow boxes
- Inclusion of habitat planting

Commonly used small-scale features include green roofs on houses and shelters, and disconnected downpipes that lead into rainwater channels.

The movement and channelling of rainwater run-off is prominent throughout the site.

Water is pumped into each courtyard and spills into planted pools through beautifully sculpted large blocks of granite.

Each courtyard contains richly planted ponds or pools.

The shopping area within the development is reached via a bridge over a planted swale that takes rainwater from the building's roofs.

Planting retention ponds

A fully functioning pond is a complex and interacting ecosystem—but it is a relatively simple thing to achieve. The balance between a beautiful pond with clear sparkling water and a murky green, soupy mess is rather fine, but thankfully a positive outcome can normally be realised in a relatively straightforward way. However, the nature of the ponds that we are discussing here, which are fed by water that has flowed over other surfaces, and are not being fed directly by rainfall or a clean stream, causes complications. This is because there are two main factors that help to achieve clear water and a wildlife-rich pond: water quality, and the amount of light reaching the pond. Water that is rich in nutrients will lead to rapid growth of the algae responsible for green cloudy water. This is a significant issue for ponds in the context of this book, because such water bodies are receiving areas for water that is flowing in from somewhere else, and by its very nature, inflowing water may be contaminated with pollutants or nutrients because of the surfaces it has flowed over. However, if water is transported and moves through swales, infiltration strips, vegetated filters and small reed beds before it reaches the pond, then these buffer areas of vegetation will have a beneficial effect on nutrient levels.

Typical section through a pond showing the main plant groups.

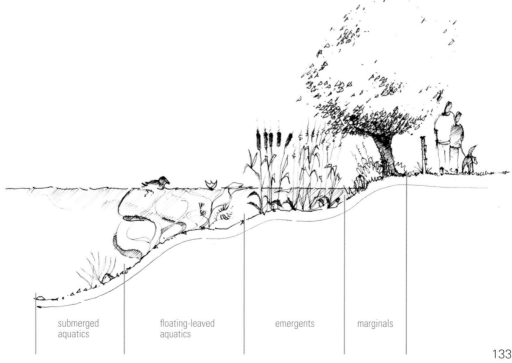

| submerged aquatics | floating-leaved aquatics | emergents | marginals |

The amount of light reaching the water body can be controlled largely through the use of planting. This is important because the higher the light levels in the water, the greater the promotion of algal growth. This isn't to say that a pond in the shade is the best option, but shading of around 50 per cent of the surface will cut down algal blooms. Typically there are several types or categories of water planting, and each have quite specific roles to play in a pond:

- Marginal plants typically grow in the permanently moist soil around the edges of water bodies. They are often the most colourful and abundant flowering plants in a wetland, and the combination of large-leaved perennials and grasses, and their flowers and seeds, provides food sources for wildlife, but also important cover for amphibians and invertebrates.

- Emergents are rooted in the mud under shallow water, and their shoots, leaves and flowers push up into the air above. Reeds and rushes are typical of this group, and they provide staging posts for aquatic invertebrates such as dragonfly larvae to climb up out of the pond when it is time for the air-living adults to leave the water.

A fully functioning pond contains a great diversity of plants and plant types, ranging from trees and shrubs on the banks through to submerged water plants.

Marginal plants, which grow best in permanently damp ground, and which will withstand periodic flooding or covering with water, are often the most showy and attractive of wetland plants and offer great potential for exploitation in rain gardens. Here flag iris (*Iris pseudacorus*) lines the edges of a lake.

- Floating-leaved aquatics live in the deeper water, but are again rooted at the base of the pond. Water lilies are the best known of this group. As well as providing shade for the pond, the leaves provide shelter for fish.

- Submerged aquatics mostly remain below the surface and are the main oxygenators of the water, but also form food sources and shelter for aquatic invertebrates and other pond life.

Emergent plants, such as reeds, grow in shallow water.

Marsh marigold or kingcup, *Caltha palustris*, is an effective marginal plant, seen here as a ribbon of yellow on the edge of a large pond.

In addition, trees and shrubs around the edge of the pond (if not on the northern edge) will provide shading for some of the day, but will also provide habitat and shelter for birds. All these different groups have a different role to play and the opportunities for wildlife, influencing light levels, and biological filtering of nutrient-enriched waters are increased if all groups are present.

Wet woodland or carr is often overlooked as a component of ponds. A degree of shading helps prevent algal blooms, and trees and shrubs are excellent for encouraging wildlife.

Bog plants need soil that does not dry out, but will not tolerate prolonged submersion under water or flooding. Often these plants have dramatic foliage as well as flowers. Here frothy white meadowsweet, *Filipendula ulmaria*, grows with the large-leaved *Inula magnifica*. Bog plants are suitable for use in rain gardens.

In the Emscher Park, Duisburg, Germany, wetland planters contain planting at different levels. At times of flood the plants will be covered. As water draws down in drier periods, the plants are still able to root down into wetter soil.

In the Fredericksburg Castle Gardens, Helsingborg, Sweden, wetland plants are grown in concrete pipes within formal pools. Regardless of changes in water level, the plant roots are always able to reach wet ground.

A similar idea used in a sustainable urban drainage scheme in Augustenborg, Sweden, uses wetland planters as an ornamental feature.

Case study Bureau of Environmental Services Water Pollution Control Laboratory, Portland, Oregon, USA

This landmark building and its surrounding landscape were designed to provide strong visual signals of water flows and movement. The landscape centres on a large water garden focused around a pond composed of two converging circles, with a striking vernacular circular stone wall housing the water outlet structure, a 30-m (100-ft) long concrete chute that brings stormwater into the pond, and rich planting that naturally facilitates sedimentation and biofiltration to ultimately produce clean water.

Water Pollution Control Laboratory, Portland, Oregon. The dramatic concrete-lined flume delivers excess stormwater into the pond.

Photographs by Tom Liptan

Water Pollution Control Laboratory, Portland, Oregon. The water exits the roof on extended scuppers and drains into the planting beneath.

Rain gardens and infiltration gardens

What are they? Shallow planted depressions.

How do they manage water? Eliminating run-off through promoting infiltration. Pollutant absorption.

Rain gardens are planted shallow basins and take the concept of stormwater landscaping to another level by enabling larger areas to be planted to both ecological and visual benefit. They have no water transport function and are designed to collect and hold stormwater run-off, allowing pollutants to settle and filter out as the water drains through vegetation and soil into the ground. Following heavy rainfall, run-off flows into the area and ponds on the surface, and gradually infiltrates through the bed of the garden. Typically, rain garden areas will infiltrate 30 per cent more water than a conventional lawn. The intention of these gardens is to promote infiltration as much as possible, however an overflow may be needed (particularly if native soils are not particularly permeable) where the basin is receiving from a large surrounding surface area, or where the planned area is close to buildings, paths or other areas of continuous use, so that any excess water can drain away rather than causing localized flooding—although in some instances this may not pose a problem. The overflow can either take the form of a surface drain, or an underdrain system. Of course, one rain garden can simply overflow into another. The Portland Manual recommends that beneath such a basin there should be at least 1 m (3 ft) of infiltratable soil or growing medium before impervious bedrock is reached, and that 30 cm (1 ft) of topsoil should form the top layer. As with other infiltration elements, rain gardens help protect streams and watercourses from pollutants carried in run-off—in a garden context this may include fertilizers and pesticides, oil and other leakages from cars, and other substances that wash off roofs and paved areas (University of Wisconsin 2003).

Planting can be diverse, incorporating all possible types of vegetation—trees, shrubs, perennials, bulbs, grasses—although typically it is dominated by perennials. Mown grass is not only maintenance-intensive, but also has much lower wildlife value, and is likely to turn

Water from the roof is channelled via a planted swale to a gentle depression contained by an earth bank. Within this richly planted depression water will be stored before being gradually released to the groundwater.

into a muddy quagmire. Because the intention of the feature is not to store water, and excess water will infiltrate or drain away, these facilities are not the same as bog gardens or rain gardens, i.e. they are not permanently wet, although a reservoir of moist soil is likely to remain at depth. Typically, bog gardens are sited alongside ponds and may be permanently saturated, although soil level may be above water level. Alternatively, they may be sited in a naturally poor-draining and wet hollow within a garden. In both cases such gardens do not function properly as rain gardens, because water does not infiltrate easily. The key criteria for planting is that it is able to withstand periodic inundation, but is not dependent on continuous flooding, and is also able to grow for much of the time in drier conditions. Although plants from

dry habitats will not tolerate these conditions, a surprisingly wide range of medium-sized trees and shrubs will withstand this regime. 'Marginal' perennials that grow naturally on the edges of ponds and lakes will also be suitable.

The concept of the rain garden is a relatively recent idea and originated in Prince George's County, Maryland, USA, in the late 1980s, where the concept of 'bioretention'—using planted areas to soak up polluted run-off water and encourage its infiltration back into the soil—was first implemented by the Maryland Environmental Protective Department in public landscape schemes. The first examples were applied in a parking lot, whereby a natural system based on planting and soil was found to be more cost-effective than a standard engineered approach. Since 1997, a collaboration with the University of Maryland has delivered performance data to support the economic and environmental case. The idea was later developed for use in private gardens and is now particularly associated with the North American midwest, in states such as Michigan and Wisconsin. Rain gardens are usually associated with the use of native plants that grow naturally in depressions or ditches. There is, however, no reason why native plants only should be used.

Many of the other elements described in this book have the characteristics of rain gardens (i.e. vegetated areas to encourage infiltration), but the main difference between them and rain gardens proper is that they tend to be relatively small, or linear, and also have functions relating to movement and transport of water as well as infiltration. Rain gardens are more substantial areas and are primarily intended to be a full stop, or at least a substantial comma, in the chain of stormwater movement, capturing as much run-off as possible.

Siting a rain garden

Rain gardens should be sited to take full advantage of run-off—this may be directly from the house or buildings to capture roof run-off, or further away to take run-off from lawned or paved areas, or from other elements in the stormwater chain. The rain garden should be positioned somewhere between where water leaves the source, and where it would naturally end up in the landscape. It would seem logical to place it at the furthest point, in the position where it would collect anyway. But this is to miss the point—the rain garden is intended to absorb as much water as possible before it gets to that point and, as mentioned previously, it is a mistake to site the garden in a poorly drained position where water already collects and ponds. When positioning close to a building, any infiltration facility should be kept at least 3 m (10 ft) from the building to avoid seepage into foundations (this rule doesn't apply to stormwater planters, which are waterproofed against the building). Siting the rain garden in a relatively flat area will make construction easier, and a position in full or partial sun will not only enable a greater diversity of planting, but will also promote evaporation of collected water.

The University of Wisconsin recommends two simple tests for whether the soil type in the chosen area is suitable (University of Wisconsin 2003). Firstly, dig a shallow hole in the ground (around 15 cm, 6 in deep) and fill it with water; if that water stays in the hole for more than 24 hours, then the site is not suitable. Secondly, take a handful of soil and dampen with a few drops of water. After kneading the soil in your fingers, squeeze it into a ball. If it remains in a ball then squeeze the soil between your forefinger and thumb, squeezing it upward into a ribbon of uniform thickness. Allow the ribbon to emerge and extend over the forefinger until it breaks from its own weight. If the soil forms a ribbon more than 2.5 cm (1 in) long before it breaks then the site is not suitable for a rain garden.

If you have to position a rain garden on a site with poor infiltration then the topsoil could be replaced by a 'rain garden mix', as recommended by the organization 'Rain Gardens of West Michigan', comprising 50 per cent grit or sharp sand, 20–30 per cent topsoil, and 20–30 per cent organic compost, and the subsoil loosened and any compaction relieved. Furthermore, an under-drain system may be used—perforated pipe in a gravel base is a simple technique.

Making a rain garden

The construction of a rain garden is relatively straightforward and simply involves lowering the level of the ground so that a rain-collecting basin is formed. Typically the depth of a rain garden is 10–20 cm (4–8 in), with gently sloping sides. This presents few problems on a level site, but on a sloping site it will be necessary to either cut and fill (remove soil from the upper part of the slope and replace at the lower) or bring in extra soil to level out the area.

The following recommendations are adapted from 'Rain Gardens: A How-to Manual for Home Gardeners', produced by the University of Wisconsin (University of Wisconsin 2003). If the rain garden is being designed to absorb 100 per cent of the run-off that might be reaching it, the total drainage area that might shed water into it must be estimated so that the approximate size of the rain garden can be calculated. If the rain garden is to take water directly from downspouts and is less than 10 m (30 ft) from the house then the approximate area of the house itself should be taken as the drainage area, if all downspouts are feeding the rain garden. If only a proportion of downspouts link directly to the rain garden then estimate the proportion of the roof area that is contributing run-off.

If the rain garden is more than 10 m (30 ft) from the building and is draining land other than roof space, then identify approximately the area that will contribute run-off, measure its length and width and multiply to find the area.

Next, estimate the infiltration ability of the soil. Sandy soils infiltrate the fastest—such soils will feel gritty and coarse if rubbed between the fingers. If the soil is silty, it will feel smooth but not sticky. Finally, clay soils will be sticky and clumpy—such soil has poor infiltration.

The total recommended area of the rain garden can be calculated by

	Rain gardens less than 10 m (30 ft) from downspout			More than 10 m (30 ft) from downspout	
	10–15 cm (3–5 in) deep	15–18 cm (6–7 in) deep	20 cm (8 in) deep	All depths	
Sandy soil	0.19	0.15	0.08	0.03	Rain garden size factors
Silty soil	0.34	0.25	0.16	0.06	
Clay soil	0.43	0.32	0.20	0.10	

multiplying the estimated run-off area by the rain garden size factor for the garden soil type (see table on page 143).

The resultant size is for a garden that will typically capture all the run-off from the drainage area. However, if any of the other storm-water capture and infiltration elements described in this book are used before water reaches the rain garden, then clearly the garden can be reduced in size. It is not necessary to site a rain garden close to a building even if is intended that the garden will take water from the building roof. A pipe can be buried to take water from a downspout further into the garden, or water can be conveyed via swales and filters to its ultimate destination in the rain garden.

In order that the rain garden captures as much water as possible, the longer side of the garden should face upslope. It is recommended that the garden should be about twice as long as it is wide, with a suggested minimum width of 3 m (10 ft).

Before digging out the rain garden, existing vegetation or turf should be removed to leave a clean surface. Over small areas, the turf can be stripped away—or the vegetation can be killed by covering for a period with black plastic, newspapers, carpets or other light-excluding materi-al, or with herbicide. You will then need to remove soil and re-profile the area to make the depression that collects water. At this stage it is sensible to thoroughly loosen the soil in the base of the garden, either with a fork, or by double digging with a spade. Incorporation of gravel or grit will help aid infiltration.

The aim of the excavation is to reduce the average height of the ground within the rain garden area by around 15 cm (6 in). To avoid having to remove large amounts of soil from the site, much of the exca-vated soil can be placed around the edge of the garden, away from the direction in which water flows into the garden, to create a slight raised lip—this will help retain water within the dug-out area. Strategically placed rocks at the point at which water flows into the garden can pro-tect against erosion during heavy flows.

Planting the rain garden

Rain gardens can be established by planting or seeding, or by a combi-nation of the two approaches. Planting is the commonly used technique,

The area to be sown was first cleared of existing vegetation or weeds. A 2.5-cm (1-in) mulch of sand was spread over the area and the prairie seed mix sown on top. A hessian mat was placed over the seeded mulch to protect from disturbance by birds and animals.

The resulting vegetation is dense and naturalistic in form.

but can be expensive over large areas, and is to some extent inefficient in that large gaps of bare soil can be left between plants in the early stages, requiring additional weeding and maintenance, even if a mulch is used. Seeding is a relatively new technique for rain gardens, although it is a well-established method for creating wildflower meadows and prairie vegetations. The advantages of sowing are that it is inexpensive to cover large areas, and can result in a fully naturalistic effect. It also results in a greater density of plants per square yard or metre, thus reducing long-term weeding requirements (although weeding input early on may be higher so that weed seedlings do not become established with the desirable plants). A combination of the two approaches gives the best of both worlds—a framework of widely spaced perennials and grasses is planted, and then the gaps and areas between the planted plants are sown.

Native or non-native? Much of the guidance on rain gardens explicitly stipulates the use of native planting. There can be sound moral arguments for the use of native planting, and for the preservation of native plant communities in gardens, but many of the so-called scientific arguments put forward about native plants being better suited to a particular climate and place than non-native plants don't really stack up. That's before we even get into the thorny discussion of what is a native species in any particular locality, and how we define nativeness. Native species are typically seen as being inherently ecological, whereas exotic (non-native) species are not, unless considered in the context of the country they hail from, in which case they immediately become ecological (Hitchmough 2003)! Non-native species are often claimed to be less well suited to local climate and environmental conditions, and therefore need greater maintenance, care and protection to ensure their survival. At the same time, exotic species are claimed to be highly invasive, dangerous, and too successful! The fact is that invasiveness has little to do with geographical origin, but is instead related to whether plants possess certain biological traits such as high seed production, effective dispersal, and low palatability to herbivores. Many native species are highly invasive and effectively dominate vegetation, leading to much reduced diversity. Similarly, many exotic species may be ideally suited or fitted to a particular environment, because they come originally from very similar environments in their own countries. There are invasive natives and there are invasive aliens. In Britain, for example, the commercially available cultivated exotic flora is in excess of 70,000 taxa, of which only a small proportion have the adverse ecological impact of invasive natives. It is also important to take into account the habitat non-native species provide for fauna in towns and cities. As with native species, non-native species differ in their value as a habitat or foraging resource but it is clear that they are very important for nature conservation in urban landscapes, as can be shown for invertebrates by the work of Owen (1991).

One of the most frequent arguments over the use of native plant species is that they support a greater number of other species (e.g. invertebrates) than non-native species, because of co-adaptation, and are therefore better for biodiversity. Generally speaking, the longer a plant has been established in a region's flora, the more insects and other invertebrates may be associated with it. This has been demonstrated most famously for trees (Kennedy and Southwood 1984). However, this

	Species in garden	Species in British Isles	Garden as % British Isles
Native flowering plants	166	c. 1500	11.1
Centipedes	7	46	15.2
Harvest spiders	10	23	43.5
Grasshoppers and crickets	3	28	10.7
Lacewings	18	55	32.7
Butterflies	21	62	33.9
Moths	263	881	29.9
Wasps	41	297	13.8
Ground beetles	28	342	8.2
Ladybirds	9	24	37.5

Numbers of species in a Leicester garden and in the UK.

(Adapted from Owen 1991)

doesn't necessarily hold true for all plant–insect relationships, and indeed many native tree species will score less well than introduced exotics. Non-native species may fill an important gap or food source at a time when natives are not in flower. While, on balance, natives do often support more feeding invertebrates, this does not merit an assumption that the conservation value of non-natives is negligible.

Some of the most relevant research to consider the value of non-natives to biodiversity comes from investigations into the habitat value of gardens. For example, Jennifer Owen has meticulously recorded the plants and animals occurring in her typical suburban garden in Leicester, England, for the past three decades (Owen 1991). A summary of her findings is shown in the table on page 147. The garden is not specifically planted with native species, and yet, just in this small plot, a significant proportion of the entire British representation of important families of plants and animals are found. She regards private gardens as the UK's largest nature reserve, and much of this arises from the diversity of non-native plants to be found in gardens.

Further evidence comes from the Biodiversity in Urban Gardens in Sheffield (BUGS) project. Seventy gardens of all types, sizes and locations were intensively sampled for their invertebrate biodiversity over a period of 3 years. Using the data gained, various comparisons were made between the amount of diversity present in the gardens and different variables, such as size of garden, location of garden, intensity of management, and whether or not the garden was planted primarily with native species. The analysis indicated that whether native species were used or not made no statistically significant difference to invertebrate biodiversity. The main factor that explained the variations in biodiversity was the structure of the vegetation—whether there are tree, shrub and herbaceous layers present, for example. The actual composition of the vegetation made less difference (Smith et al. 2005). In other words, the key factor in urban biodiversity is not the geographical origin of the plant species used (although this is critical for some fauna), but rather taxonomic diversity and spatial complexity of planting and landscape spaces. In short, lots of different plant species arranged in as many layers as possible. Finally, insistence on natives-only on sites that do not have designated nature conservation value has unfortunate connotations in a multi-cultural society (Wilkinson 2001). So, while the use of native species is desirable for many reasons, the opposite argument—that non-native species have no or much reduced value for promoting bio-diversity—does not necessarily hold true, depending on context.

The point of this discussion is that we can make our own choices as to what we may wish to plant in a rain garden, and do not have to be tied down to a natives-only rule. The plant directory in Section Three gives a full account of the range of plants that may be used in rain garden landscapes.

Maintenance of rain gardens

In keeping with the more ecological ethic, rain gardens should not require intensive maintenance. Because they work partly through the actions of plants, they should be plant-dominated, and large areas of bare ground are not useful. Two main tasks will be necessary. Firstly, aggressive weeds should be eliminated at the outset. Weeding is essential in the first few months as plants are becoming established. Once a good vegetation cover has been achieved, the annual need for weeding will be reduced. Secondly, all the above-ground growth will need to be cut back. Most naturalistic gardeners now prefer to leave growth standing through the winter—not only are the dried-out stems and seed heads attractive, but they also provide winter food sources for seed-eating birds, and some shelter for over-wintering invertebrates. By cutting back the dead stems in late winter and early spring, the garden is cleared for the new season. If the cut material is very dry it can be simply broken up and applied as a mulch to the soil, or it can be disposed of by other means. There will be no need to fertilize the garden. Irrigation is also out of bounds: it goes against the ecological spirit of this book. But some watering in of plants in the first few months after planting may be essential to ensure success. In exceptionally dry periods some watering may be necessary—this can use grey water from the house (see Section One) if for short periods of time.

Case study Buckman Heights, Portland, Oregon, USA

Two rectangular infiltration gardens form the centre of a courtyard and take run-off from the rooftops and surrounding paved areas. Plants such as *Mahonia aquifolium*, *Iris sibirica* and *Astilbe* form the bulk of the planting—all these are tolerant of wet conditions, but will also grow well in normal fertile soil so long as it does not dry out completely.

Run-off flows into the basin and soaks into the soil, except for large rainfall events that flow out of the overflow—the overflow pipe is set to allow 22 cm (9 in) of collected water. The experience of this site suggests that infiltration works very well—in fact the height of the overflow pipe could be increased (Liptan 2002). Annual clearance of material such as leaves washed down from roofs is required to maintain good drainage flow.

Buckman Heights rain garden. The central planted beds receive excess run-off from the paved surfaces and building roofs.

Photograph by Tom Liptan

Urban Water Works, Portland, Oregon, USA

Urban Water Works began in 1999, having been inspired by the work of the designer Betsy Damon in creating a 'living water garden' in Chengdu, China, which was planned around the art, science and educational value of water in the urban environment. Urban Waterworks seeks to create a network of community-based projects in the city that make water visible within neighbourhoods, based upon the way water might have moved through the landscape prior to built development. They work very closely with local volunteer groups and their projects typically involve transformation of school grounds from barren tarmac deserts into green and vibrant places. Two well-established projects are the DaVinci Water Garden and the Astor Water Garden.

DaVinci Water Garden

This water garden at the DaVinci Arts Middle School in Portland was initiated in 2000 as the result of an interdisciplinary exploration of water that was undertaken as an educational exercise by the school. The aim of the garden is to manage run-off from adjacent impervious surfaces, as well as being used as an outdoor laboratory for the school, and a community facility.

The garden was constructed on a former tennis court. The site was excavated and the site remodelled to contain a lined pond, a vegetated swale, a rain garden and two rainwater harvesting cisterns. The cisterns store 19,000 litres (5000 gallons) of water and supply the pond with water, and much of the irrigation needs of the garden. The garden absorbs run-off from around 450 m² (5000 ft²) of surrounding roofs and hard surfaces. Around half of this amount is directed into the pond, and half into the cisterns. Overflow from the pond is directed into the swale, which drains into the rain garden. The upper third of the swale is lined with a waterproof membrane so that during heavy rain a visible stream forms. The lower two-thirds are not lined, to encourage infiltration. In addition, a further swale drains water from an adjacent car park into the main swale. The rain garden is planted with native grasses, rushes and flowering plants. Overflow from the rain garden drains into an adjacent trench alongside a playing field. It has been calculated that the garden prevents around 1.2 million litres (315,000 gallons) of water from entering the main stormwater sewers that drain the site.

The DaVinci Water Garden one year after planting. The adjacent car park, school and water cistern.

The path leading into the swale.

Photographs by Brock Dolman

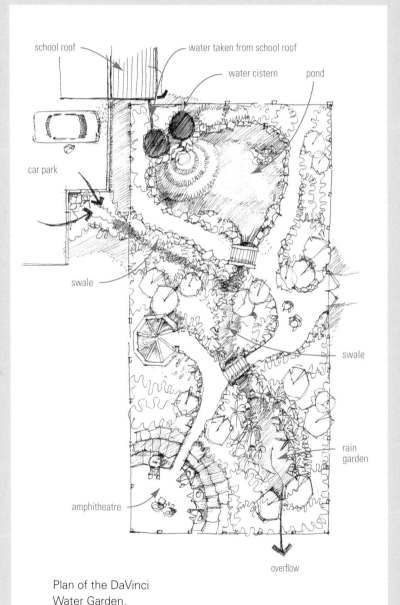

Plan of the DaVinci Water Garden.

Astor Water Garden

Elementary students, educators and the community helped design and build this vibrant learning garden, which helps clean and filter stormwater run-off from what was formerly 800 m² (8,500 ft²) of asphalt. Planted in spring 2005 by volunteers and with contributions from many professionals and businesses, the Astor Water Garden combines art and environmental education with community design.

The existing site consisted of 800 m² (8,500 ft²) of asphalt that was visually monotonous, very hot in summer and bleak in winter, and shed all water that fell on it. The lines of the garden were painted onto the asphalt before excavation. All the asphalt was removed to expose the soil underneath. New topography and landform was created using imported soil and the water collecting and receiving areas dug out. Volunteers, teachers, families and children helped to plant the rain garden.

The main shapes of the garden were marked on the tarmac with spray paint.

Photographs by Erin Middleton

Before work started, the site was a barren tarmac yard that shed all water that fell on it.

The tarmac was removed to expose the underlying soil.

New topography and landform was created using imported soil and the water collecting and receiving areas dug out.

Volunteers, teachers, families and children helped to plant the rain garden.

The completed site after planting.

Photographs by
Erin Middleton

Swimming ponds

What are they? Naturally-cleansed water bodies, with clear, uncontaminated water that can be used as swimming pools.

How do they manage water? Swimming ponds are fed by rainwater and utilize the water-cleansing properties of wetlands for their functioning.

The first swimming ponds were built in Austria in the mid-1980s as an ecological alternative to chemically treated and cleansed swimming pools. Now there are over 20,000 private swimming ponds in Austria, 8000 in Germany and 1500 in Switzerland, as well as several hundred local authority public natural swimming pools (Littlewood 2006). They are a chemical-free combination of a swimming pool and a rain garden. They can be breathtakingly beautiful, combining sumptuous water plantings with the crispness of modern materials and geometric or freeform designs, and provide all the benefits of any wildlife pool. The pool water is pumped through a planted 'regeneration zone', established in shingle or gravel, rather than topsoil. Plants must take their nutrients from the water, thereby cleansing it, and producing crystal-clear water for the open 'swimming zone': the area of open water that is kept clear of vegetation for free and easy swimming, and is flat-bottomed, with typical depths of 1.5–2.0 m (5–6.5 ft). In a garden setting they provide great horticultural and habitat potential.

The reasons for including them in this book are two-fold. Firstly they make use of plants and vegetation to physically and chemically cleanse water. Secondly they are systems that do not use mains water—municipal water from the pipe contains nutrient loading that if continuously added to the pond will promote unwanted algal growth. Instead, swimming ponds tend to be rain-fed—for example being topped up by harvested rainwater. They therefore fulfil some core requirements of a bioretention system.

The key difference between the swimming pond and most of the other features we have discussed is the crucial need to maintain a very low nutrient status in the water. As soon as nutrient levels rise, algal growth is promoted and the water becomes cloudy, green and unpleasant

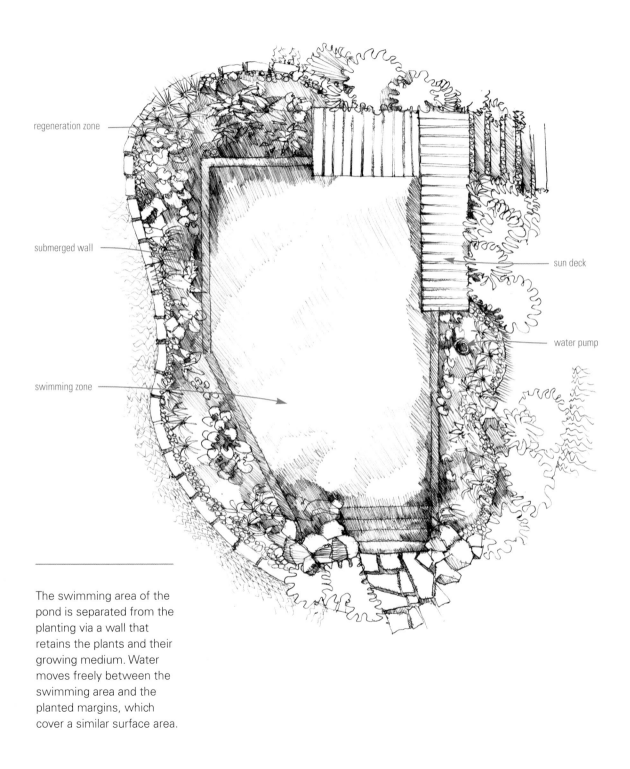

regeneration zone

submerged wall

swimming zone

sun deck

water pump

The swimming area of the pond is separated from the planting via a wall that retains the plants and their growing medium. Water moves freely between the swimming area and the planted margins, which cover a similar surface area.

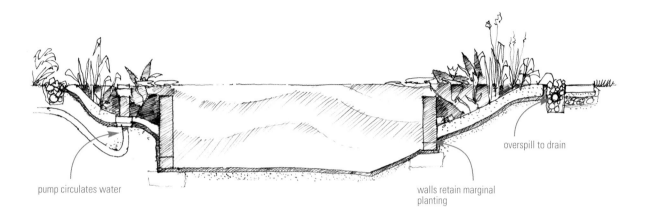

pump circulates water

overspill to drain

walls retain marginal planting

to swim in. The means by which water is cleansed in the regeneration zone is complex but is mainly linked to the action of micro-organisms on plant roots and substrate particles that break down impurities, together with the oxygenating properties of plant roots. The substrate and plants also help to filter out particles and debris. However, unlike the other examples in the book that tend to rely on normal soils to support planting, swimming ponds do not use soil at all, but an inert, low-nutrient substrate such as gravel. In addition, ultraviolet filters can be fitted to remove any potentially harmful bacteria from the water. The ponds are constructed in a similar way to a normal garden pond, using a heavy-duty liner. No substrate or other materials are placed on top of the liner in the swimming zone.

Swimming ponds differ in their complexity, from relatively simple systems where the swimming zone and the regeneration zone are found in the same pond, to more complex forms where the two zones may be separated into different ponds, and they also differ in the degree of planting, with some types consisting mainly of gravelly substrate rather than lush vegetation.

One of the most attractive features of swimming ponds is that they often combine rich and complex naturalistic planting with a strong geometric framework. Normally naturalistic water planting is associated with naturalistic or informally shaped ponds, but this doesn't have to be the case.

The water is circulated between the planted margins and the swimming zone with a water pump.

Case study The Kircher Pond, Germany

Professor Wolfram Kircher, a leading German researcher into perennial plants and aquatic planting, designed and constructed a swimming pond in his own garden eight years ago to address a major problem he had identified through observing some of the longest-running swimming ponds in Germany. "There was a conflict between needing to maintain very low levels of nutrients in the water to prevent algal blooms and murkiness, and being able to grow the sorts of plants that were being used in the regeneration zone." These common wetland and marginal plants were usually vigorous and needed relatively productive water and soils for them to grow effectively. Wolfram noticed that these plants tended to become very nutrient-stressed over a period of several years, particularly after hot summers. However, they were usually the plants that were easily available from aquatic nurseries and garden centres. "I decided to look to nature for alternatives", says Wolfram, "and to get my inspiration from naturally low-nutrient wetland habitats, such as bogs and fens."

In Wolfram's swimming pond the regeneration zone is physically

Wolfram Kircher's pond is in constant use throughout the summer months by his family. When not in use it is a beautiful feature in itself.

separated from the swimming zone into different areas, with the regeneration zone taking the form of a meandering, running stream. "The regeneration zone is most effective when it is narrowly shaped or even takes the form of a stream, because then water movement brings about improved nutrient uptake by the vegetation, thereby improving the effectiveness of the water-clearing function." Separating the two areas also means that a greater range of planting can be used.

The water is circulated by a pump concealed beneath a timber bridge that spans the pond. Water emerges through a 'spring' and tumbles along a rill through paving before it enters the regeneration zone. This area is intensely planted with strongly growing emergent species such as irises, tufted sedge (*Carex elata*), water mint (*Mentha aquatica*), meadowsweet (*Filipendula ulmaria*) and purple loosestrife (*Lythrum salicaria*). Amongst these dominants, less vigorous species are interspersed: marsh marigold (*Caltha palustris*), ragged robin (*Lychnis flos-cuculi*) and devil's bit scabious (*Succisa pratensis*).

While the aim of the regeneration zone is for the vigorous plants to extract and reduce the nutrient loading of the water, the aim for the swimming zone is to maintain crystal-clear, low-nutrient water. Wolfram has planted a very different type of vegetation along the edges of this zone: an artificial bog vegetation with an exciting mix of grasses, insectivorous plants, dwarf shrubs (heaths and bilberries), orchids, and delicate wetland species. Here sphagnum moss has been successfully established as ground cover from cuttings pushed into the wet substrate surface. Because there is no water-cleansing function, the vegetation here does not have to be totally submerged. Fluffy white cotton grasses (*Eriophorum vaginatum*) make a dramatic show in June, interspersed with dramatic purple spikes of marsh orchids (*Dactylorrhiza* hybrids), which were established from seed, and now self-sow happily. A distinctive late-flowering component is the wetland aster, *Aster puniceus*, which flowers from late July through to October with purple flowers on a low spreading, creeping and weaving plant. Wolfram has specialized in two different groups of plants here: insectivorous plants and semi-parasitic plants. The insectivorous plants thrive in the nitrogen-deficient conditions and include sundews (*Drosera* species) and pitcher plants (e.g. *Sarracenia purpurea*).

Water is pumped into the swimming zone through the lushly planted regeneration zone.

Water flowing through this channel is aerated before it passes into the regeneration zone.

DESIGNING YOUR RAIN GARDEN

So far we have discussed each of the components that might be included within a rain garden and, through case studies, explored how these techniques have been applied in a wide range of different settings. How do you go about applying these techniques to your own garden or landscape? Before answering this question you need to think long and hard about what your own aspirations for your garden are, and how a rain garden may help achieve these objectives. Remember, a rain garden isn't just a functional solution, it also creates different opportunities for incorporating new planting, wildlife habitat and play. For example you may not want the cost and upkeep of a pond but still want to create a rain garden. You might also wish to adopt an approach that lets you implement a rain garden in stages or which deals with an isolated building within the garden. Whichever approach you take it is a positive step towards sustainable water management.

SITE SURVEY

By knowing and understanding your site you are already halfway towards creating a successful rain garden: in many ways once you have come to know the site and the way that water works within it, and could potentially work in the future, then the garden itself will start to make suggestions to you about the most appropriate way in which it might be altered to accommodate rain garden features.

Rainwater catchment and drainage

The first step in developing your rain garden is to identify all of the different impermeable surfaces where rainwater falls and how this is discharged. Is the water drained to the local drainage network or dispersed within the garden? The largest surface will typically be the roof of the house and possibly other garden buildings, and paved or sealed surfaces. In areas with particularly high rainfall or on sites where the land slopes steeply towards a property there may also be land drains installed that connect with the drainage network. It may be possible to locate these by inspecting the drainage manholes.

At this point it is advisable to draw a sketch plan showing each of these surfaces and where they drain to. For example the roof of a building may have two pitches, one that drains to the front of the property and the other to the rear. Both will typically drain into the drainage network. It would also be helpful to calculate the relative size of these areas. In order to calculate how much rainfall this surface will generate you will need to find out your local rainfall data, which can then be multiplied by the surface area to give a volume in litres/gallons. Within the section on water butts we give an example of the amount of water that would be expected to flow from a roof of 110 m^2 (1180 ft^2). At this point it is important to remember that rain gardens are not an either/or solution. If you wish you can restrict your rain garden to a shed or greenhouse, or alternatively use a downpipe connector that automatically shuts off when your water butts have been filled.

Topography

You need to assess the natural topography and hence drainage of the garden—water naturally flows to the lowest point. Understanding the topography will be essential in determining which elements of the rainwater chain are required and where these should be located. For example if there is a natural slope across a garden, a swale may be used to gather this water before directing it to the 'rain garden' or 'pond'. You could add this information to your plan using arrows to show the direction of fall and the lowest and highest points of the garden.

Although some remodelling of the garden will be necessary to create swales, ponds and rain gardens it is best if this works with the existing topography. Excessive landform change is not desirable and may be costly and result in removal of established vegetation and damage to soil structure that may impede drainage and infiltration.

Soils and permeability

If you have been gardening your plot for some time you will already be very familiar with the quality of the soil and how it may vary within the garden. In determining where, for example, ponds or infiltration devices should be located, an important factor will be the relative capacity of the soil to absorb water. One approach to assessing how well the soil drains is to survey the garden shortly after a heavy downpour and then again later in the day, to assess how quickly the water has dispersed. You may even wish to excavate some trial pits, which may reveal where the natural water table is and how quickly the water disperses within the soil.

Vegetation

Vegetation that is already established within the garden (and potentially in neighbours' gardens) should be considered throughout the design process. Roots may be damaged in the process of excavating infiltration devices and groundwater conditions may change and not be suitable for certain species. Heavy shade from existing trees may limit plant growth. Again it is advisable to map existing vegetation onto your plan in order to ensure that you take account of it while designing your rain garden.

Services

You should also map all existing services that may run through the garden, including electric, gas, water and sewage/drainage. Unfortunately these are not always clear—inspection covers may give some indication of where they are located but there are plenty of occasions where a water pipe or electricity cable are unearthed in the process of excavating a garden. In some cases the only option is to proceed with caution.

The plan and section
record. Impermeable
surfaces: roof,
garage, drive, patio
and greenhouse.
Location of mature
planting. Approximate
levels and drainage.

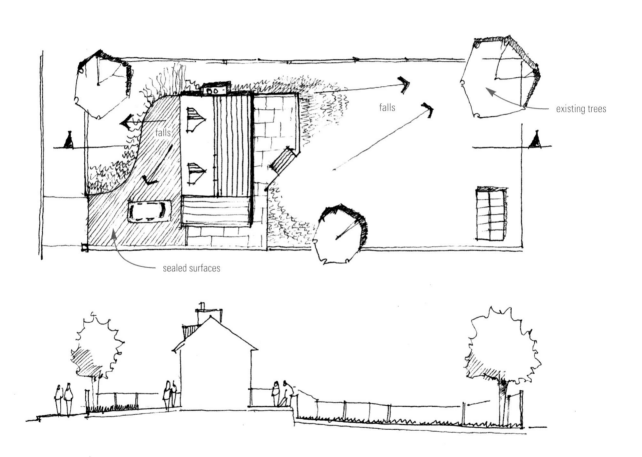

falls

falls

existing trees

sealed surfaces

Context

A rain garden is not a closed system. Rain enters and water leaves. This water might leave quite harmlessly into groundwater or an adjacent stream. We must however be aware of the possible ramifications if the system overflows or if by our actions we have a significant impact on the drainage of an adjacent property. If there is any doubt then restrict the amount of water entering the rain garden by controlling the amount of water leaving the downpipe and entering the rain garden. The other alternative is to incorporate an overspill pipe that is connected to the main drainage network.

SETTING OUT A DESIGN

You should now be in a position to begin designing the layout of your rain garden. You have a brief that establishes the qualities and components that you would like it to incorporate and a site plan that identifies the different factors that need to be taken into account and that also guide your decision-making concerning where different elements might best be located and how they can be brought together.

It would be helpful to draw your plan at a recognizable scale. This will then help you to assess the approximate size of the various components and their water storage capacity. It will also help you to consider other aspects of the design including new planting and how surplus topsoil from ponds and swales might be used within the design to create berms and other features. Once the main structure has been set out it is then possible to begin to consider the finer details of the design, which will elevate the scheme from being a practical solution to a rich and pleasurable design. The design on the next page shows how the various components of the stormwater chain might be applied.

Excess water from the pond runs into a planted swale, which eventually runs into a rain garden. Water from the greenhouse is directed into an internal water butt. In winter this is redirected to two large external water butts. Overspill is allowed to soak into the adjacent vegetable plot. The lawn area has been reduced and planting increased. This will improve water capture and evapotranspiration.

The new design aims to create more opportunities for capturing rainwater and then slowly releasing it or using it within the garden.

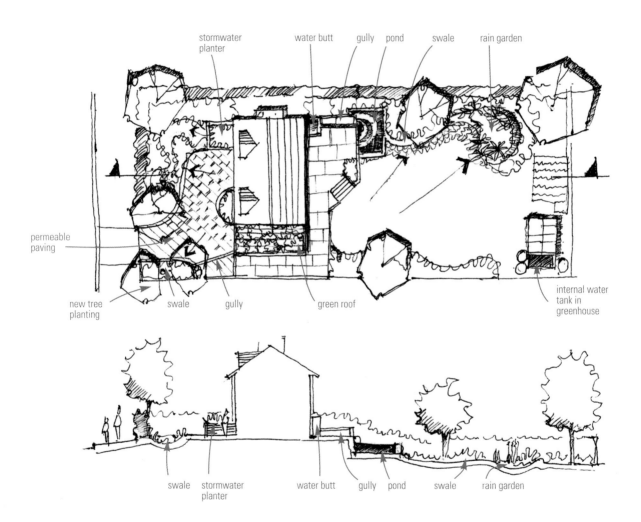

stormwater planter

water butt gully pond swale rain garden

permeable paving

new tree planting swale gully green roof

internal water tank in greenhouse

swale stormwater planter water butt gully pond swale rain garden

Front garden. A stormwater planter captures water from the roof downpipe. When full it overspills into a swale that transports any excess to a small rain garden. The tarmaced drive has been reduced in area and replaced with a permeable paving block. Excess water from the drive is directed into planted swales. If these should reach capacity they will overspill into a drain connected to the drainage network. The garage roof has been replaced with a green roof. Excess run-off is guided through a channel set in the paving into the planted swale. Additional tree and shrub planting will reduce the amount of water falling on paved surfaces and increase evapotranspiration.

Rear garden. A large water butt captures water from the roof downpipe. The water butt overspills into a channel that cuts across the patio and feeds a small formal pond. Excess water from the pond spills into a swale which then leads to a rain garden. Here the water will gradually infiltrate into the soil. Water from the greenhouse roof is directed to an internal tank to regulate the temperature of the greenhouse and for watering plants. Excess water can be directed to water butts for use in the garden.

Floating-leaved aquatic
plants such as water lilies
are the quintessential
pond plants, and have an
important role in providing
shade and cover to other
life in the pond.

SECTION 3

Plant directory

The list of potential plants for rain gardens and other rainwater infiltration features is enormous. While it is relatively simple to describe the sorts of plants that might be found in a typical water body—ranging from the typical submerged aquatic plants (the 'oxygenators'), the floating-leaved plants such as water lilies, through to the plants of a pond or lake margin that grow in areas of shallow water—plants for rain gardens are much less easy to typify. This is because a rain garden is neither wet nor dry, but periodically swings between the two. What we can say is that a rain garden in most temperate climates is very unlikely to dry out completely for extended periods of time—even in long drought periods it is likely that there will be some soil moisture reserve at lower levels. Similarly, it is unlikely that the rain garden will suffer prolonged periods where plants are submerged under water—the essence of a rain garden is that excess water will drain away, rather than contributing to the formation of a permanent pond. Typical rain garden plants will therefore have an intermediate strategy—usually being found in situations around water bodies or in areas of moist soil, or from habitats (such as prairies or hay meadows) that are subjected to, and soak up, significant amounts of rainfall during part of the year.

In choosing appropriate plants we are helped by some general biological facts. Plants that typically grow in regularly moist habitats are often able to survive pretty well in drier soils in a garden situation where maintenance can control aggressive weeds, so long as the soil is reasonably fertile. However, the opposite is usually not true—plants that are adapted to dry conditions tend not to cope with waterlogged soils or flooding. Immediately we can begin to narrow our plant choices down. But we must not forget the dryland plants, because any rain garden consists of a gradient of moisture levels, and it is likely that the margins will never get flooded or wet. Some drought tolerance is therefore desirable. In most cases, plants in this situation will be able to send roots down to damper soil, but there may be some situations where the surroundings of a rain garden need to be planted with plants that visually appear to fit, but that are not wetland plants at all. For example, rain gardens created on free-draining sandy soils will dry out relatively rapidly. Plants tolerant of wetter conditions tend to have larger leaves and bolder foliage (where water isn't in continuous short demand, plants don't need to adopt adaptations to preventing water loss such as small leaves, narrow leaves or silver or grey hairy foliage): sometimes we will need plants that look like wetland plants but don't need moist soil in which to grow.

Potentially, therefore, we have a huge number of plants from which to choose, and with which to experiment. Rain garden plants tend to be thought of as herbaceous perennials and grasses, but we need to consider trees and shrubs too, particularly if we have a large area to deal with. Annuals can also be successful, but bulbs may not do so well because they tend to rot away if soil is continually wet.

We have already discussed the issue of natives vs non-natives in rain garden planting, and pointed out that this is a contentious issue. Virtually all references to rain gardens advise the use of American native species, and wherever possible, native plants of the particular region. Because the rain garden is essentially an American concept this isn't surprising. But in a book that has a wide readership across continents, this imposes a limiting factor on plant choice. The fact is that there is a whole array of plants from across the world that are suited to rain garden use in different countries, and in all countries. There is currently a great interest in many European countries in the visual splendour of the North American prairie and in adapting colourful prairie planting for garden use—particularly where the native flora may not be so grand or flower so late in the season. We are therefore fortunate that the majority of species regularly recommended for rain garden use in North America will also be seized upon by European gardeners as being highly desirable.

In the following section we present a series of suggestions of suitable plants for the situations we have described in this book. We have arranged the plants by categories of use, rather than by geographical origin, but have indicated, in broad terms, their native ranges. This can only be a selection and is not in any way intended to be comprehensive. Indeed, the best way to find out what succeeds best in your area is to give it a go—some plants may fail but through a process of selection you will arrive at those that are most successful for you. In the lists that follow, where one member of a genus is indicated (a particular species of willow for example), it is worth considering other members of that genus because in broad terms they are likely to be adapted to similar conditions.

The tables that follow give a broad indication of likely preferences of the listed plants in terms of their tolerance of soil moisture. Only plants for a rain garden proper are included here. There are a great number of books that cover pond, pool and water gardening and so we have not included aquatic plants here. Similarly, dryland plants are omitted from these lists. The tables indicate the geographical origin of the plants. Where the tables indicate that a species grows in sun, it is important to state that this assumes adequate moisture supply.

The moisture tolerance of the listed species is shown in four categories. These are guideline categories only and are intended to provide an approximate spectrum of requirements. The four categories are explained below:

- **Wet:** Site constantly waterlogged with long continuous periods of standing surface water. Includes swamp and marsh conditions.

- **Moist:** Soil constantly moist. Plants tolerant of longer periods of flooding.

- **Mesic:** Soil neither excessively wet nor excessively dry. Plants tolerant of brief periods of flooding.

- **Dry:** These plants will tolerate extended dry periods.

You can use these categories to judge the appropriate placing of plants within the rain garden or bioretention feature. Because the categories relate to the relative frequency of flooding and standing water, this will also depend on the amount and intensity of rainfall, and also the retentiveness of the soil.

HERBACEOUS PLANTS

Name	Common name	Origin	Height	Colour	Bloom time	Wet	Moist	Mesic	Dry	Notes
Allium cernuum	Nodding wild onion	N. Am.	0.5 m	Pink/white	Jul–Aug					Sun
Amorpha canescens	Leadplant	N. Am.	1.0 m	Purple	Jun–Aug		●	●		Sun
Amsonia tabernae-montana	Blue star	N. Am.	0.5 m	Blue	Apr–May		●			Sun
Aquilegia canadensis	Columbine	N. Am.	0.6 m	Red/yellow	May–Jun			●		Shade
Aruncus dioicus	Goatsbeard	Eur.	1.5 m	White	Jul–Aug		●	●		Sun/shade
Asclepias syriaca	Common milkweed	N. Am.	1.0 m	Pink	Jun–Aug			●		Sun
Asclepias tuberosa	Butterfly weed	N. Am.	0.6 m	Orange	Jul–Sept				●	Sun
Aster azureus	Sky blue aster	N. Am.	0.2 m	Blue	Aug–Oct			●	●	Sun
Aster lanceolatus	Smooth aster	N. Am.	1.0 m	White	Aug–Oct		●	●		Sun
Aster novae-angliae	New England aster	N. Am.	1.0 m	Blue	Aug–Oct		●	●	●	Sun
Aster novii-belgii	New York aster	N. Am.	1.0 m	Blue	Jul–Oct		●	●		Sun
Aster puniceus	Purple-stemmed aster	N. Am.	1.5 m	Blue	Aug–Sep	●	●	●		Sun
Aster umbellatus	Flat-topped white aster	N. Am.	1.0 m	White	Aug–Oct		●	●		Sun
Astilbe cv	False spirea	Asia	0.6–1.0 m	Various from white to purple	Jul–Aug		●	●		Sun
Caltha palustris	Marsh marigold	Eur. N. Am. Asia	30 cm	Yellow	Apr–May	●	●			Sun/shade
Cardamine armara		Eur.	40 cm	White	Apr–Jun		●	●		Sun

HERBACEOUS PLANTS

Name	Common name	Origin	Height	Colour	Bloom time	Wet	Moist	Mesic	Dry	Notes
Cardamine pratensis	Cuckoo flower	Eur.	30 cm	Lilac	Apr–May		●	●		Sun
Chelone glabra	Turtlehead	N. Am.	1.0 m	White	Jul–Sept	●	●			Sun
Dryopteris cristata	Crested woodfern	N. Am.	1.0 m				●	●		Shade
Echinacea pallida	Pale purple coneflower	N. Am.	1.0 m	Purple	Jun–Jul			●	●	Sun
Echinacea purpurea	Purple coneflower	N. Am	1.0 m	Purple	Jul–Oct			●		Sun
Equisetum hyemale	Horsetail	N. Am. Eur.	60 cm				●	●		Sun/partial shade
Eupatorium cannabinum	Hemp agrimony	Eur.	1.0 m	Pink	Jul–Aug		●	●		
Eupatorium fistulosum	Joe Pye weed	N. Am.	1.5 m	Pink	Jul–Sep			●		Sun/partial shade
Eupatorium maculatum	Spotted Joe Pye weed	N. Am.	1.5 m	Purple	Jul–Oct	●	●			Sun/partial shade
Eupatorium perfoliatum	Boneset	N. Am.	2.0 m	White	Jul–Oct		●	●		Sun
Eupatorium purpureum	Savannah Joe Pye weed	N. Am.	2.0 m	Pink	Jul–Aug		●	●		Sun
Filipendula ulmaria	Meadow-sweet	Eur.	1.0 m	White	Jul–Aug		●	●		Sun
Filipendula rubra 'Venusta'		N. Am.	2.0 m	Pink	Jul–Sep		●	●		Sun
Fragaria virginiana	Wild strawberry	N. Am.	0.2 m	White	Apr–Jun			●	●	Sun
Fritillaria meleagris	Snake's head fritillary	Eur.	0.2 m	Pink	Apr–May		●			Sun
Geum rivale	Water avens	Eur.	30 cm	Pink	May–Jun		●			
Geum triflorum	Prairie smoke	N. Am.	0.2 m	Pink	Apr–Jun			●	●	Sun

HERBACEOUS PLANTS

Name	Common name	Origin	Height	Colour	Bloom time	Wet	Moist	Mesic	Dry	Notes
Gladiolus palustris		Eur.	0.5 m	Pink	Jul–Aug		●	●		Sun
Helenium autumnale	Sneeze-weed	N. Am.	1.0 m	Yellow	Aug–Oct	●	●			Sun
Helianthus giganteus	Giant sunflower	N. Am.	2.0 m	Yellow	Aug–Sep	●	●			Sun
Helianthus laetiflorus	Showy sunflower	N. Am.	1.0 m	Yellow	Aug–Sep			●		Sun
Helianthus mollis	Downy sunflower	N. Am.	1.0 m	Yellow	Aug–Sep			●	●	Sun
Heliopsis helianthoides	False sunflower	N. Am.	1.5 m	Yellow	Jul–Sep			●	●	Sun
Heliopsis occidentalis	Ox-eye sunflower	N. Am.	1.0 m	Yellow	Aug–Sep			●	●	Sun
Inula magnifica		Asia	2.0 m	Yellow	Jul–Sep		●	●		
Inula racemosa 'Sonnespeer'		Asia	2.0 m	Yellow	Jul–Sep		●	●		
Iris kaempferi		Asia	70 cm	Purple	Jul–Aug	●	●			
Iris pseudacorus	Yellow flag	Eur.	1.0 m	Yellow	Jun–Jul	●	●	●		
Iris shrevei	Wild iris	?	0.5 m	Purple	May–Jul	●	●	●		Sun
Iris versicolor	Swamp iris	N. Am.	0.5 m	Blue	May–Jun	●	●			Sun/partial shade
Leucojum aestivum	Summer snowflake	Eur.	0.5 m	White	May–Jun		●			
Liatris spicata	Marsh blazing star	N. Am.	1.0 m	Blue	Jul–Sep		●	●		Sun
Ligularia dentata		Asia	1.0 m	Yellow	Jul–Aug		●			Shade
Ligularia przewalskii		Asia	1.5 m	Yellow	Jul–Aug		●			
Lobelia cardinalis	Cardinal flower	N. Am.	0.6 m	Red	Jul–Sep	●	●			Sun

HERBACEOUS PLANTS

Name	Common name	Origin	Height	Colour	Bloom time	Wet	Moist	Mesic	Dry	Notes
Lobelia siphilitica	Great blue lobelia	N. Am.	1.0 m	Blue	Aug–Sep	●	●			Sun
Lychnis flos-cuculi	Ragged robin	Eur.	40 cm	Pink	May–Jun		●	●		Sun
Lysimachia clethroides	Gooseneck	N. Am.	80 cm	White	Aug–Sep		●	●		
Lysimachia nummularia	Golden creeping Jenny	Eur.	5 cm	Yellow	Jun–Jul		●	●		Partial shade
Lysimachia punctata	Yellow loosestrife	Eur.	90 cm	Yellow	Jun–Jul		●	●		
Lythrum salicaria	Purple loosestrife	Eur.	80 cm	Purple	Jul–Aug	●	●	●		Can be very invasive
Lythrum virgatum	Purple loosestrife	Eur.	1.0 m	Purple	Jul–Aug		●	●		
Matteuccia pensylvanica	Ostrich fern	N. Am.	70 cm				●	●		Partial shade
Matteuccia struthiopteris	Shuttlecock fern	Eur.	70 cm				●	●		Partial shade
Mentha aquatica	Water mint	Eur.	50 cm	Purple	Jun–Jul	●	●			Sun
Mimulus ringens	Monkey flower	N. Am.	0.5 m	White	Jun–Sep	●	●			Sun
Monarda didyma	Beebalm	N. Am.	1.2 m	Red	Jul–Sep		●	●		Sun/partial shade
Monarda fistulosa	Bergamot	N. Am.	1.0 m	Lavender	Jul–Aug		●	●		Sun
Myosotis palustris	Water forget-me-not	Eur	40 cm	Blue	May–Jul	●	●			Sun/partial shade
Oenothera biennis	Evening primrose	N. Am.	1.0 m	Yellow	Jul–Oct		●	●		Sun
Osmunda cinnamomea	Cinnamon fern	N. Am.	0.5 m				●	●		Sun/ partial shade
Osmunda regalis	Royal fern	Eur.	1.5 m				●	●		

HERBACEOUS PLANTS

Name	Common name	Origin	Height	Colour	Bloom time	Wet	Moist	Mesic	Dry	Notes
Penstemon digitalis	Bearded foxglove	N. Am.	1.0 m	White	May–Jul		●	●		Sun
Persicaria amplexicaule cvs		Asia	1.0 m	Red/pink/white	Aug–Oct		●	●	●	Sun/partial shade
Persicaria bistorta	Bistort	Eur.	80 cm	Pink	May–Jun		●	●	●	Sun/partial shade
Petalo-sternum purpureum	Purple prairie clover	N. Am.	0.5 m	Purple	Jun–Aug		●	●		Sun
Petasites hybridus	Butterbur	Eur.	0.5 m	Pink	Mar–Apr		●	●		Sun/ partial shade
Phlox divaricata	Woodland phlox	N. Am.	0.2 m	Blue	Apr–Jun			●	●	Shade
Phlox pilosa	Downy prairie phlox	N. Am.	0.2 m	Pink	Apr–Jun		●	●		Sun
Physostegia virginiana	Obedient plant	N. Am.	1.0 m	Pink	Jun–Sep		●	●		Sun/partial shade
Primula beesiana	Candelabra primula	Asia	60 cm	Lilac	Jun		●	●		Sun/partial shade
Primula bulleyana	Candelabra primula	Asia	70 cm	Orange	Jun		●	●		Sun/partial shade
Primula florindae	Candelabra primula	Asia	60 cm	Yellow	Jun		●	●		Sun
Primula japonica	Candelabra primula	Asia	50 cm	Purple	Jun		●	●		Sun
Primula vulgaris	Primrose	Eur.	10 cm	Yellow	Mar–Apr		●	●		Shade
Pycnath-emum virginianum	Mountain mint	N. Am.	1.0 m	White	Jul–Sep		●	●		Sun/partial shade
Ratibida pinnata	Yellow coneflower	N. Am.	1.2 m	Yellow	Jun–Aug		●	●	●	Sun
Rheum palmatum	Rhubarb	Asia	2.5 m	Pink	Jun–Jul		●	●		Sun
Rodgersia pinnata		Asia	90 cm	Cream	Jul–Aug		●			

HERBACEOUS PLANTS

Name	Common name	Origin	Height	Colour	Bloom time	Wet	Moist	Mesic	Dry	Notes
Rudbeckia fulgida	Black-eyed Susan	N. Am.	1.0 m	Yellow	Jul–Oct		●	●		Sun/partial shade
Rudbeckia laciniata	Tall coneflower	N. Am.	1.5 m	Yellow	Jul–Sep		●	●		Sun/partial shade
Rudbeckia subtomentosa	Sweet black-eyed Susan	N. Am.	1.0 m	Yellow	Aug–Sep			●	●	Sun
Rudbeckia triloba	Brown-eyed Susan	N. Am.	1.0 m	Yellow	Jul–Oct				●	Shade
Schizostylis coccinea cvs	Kaffir lily	S. Afr.	60 cm	Pink	Oct–Nov		●	●		Sun
Silphium laciniatum	Compass plant	N. Am.	2.0 m	Yellow	Jun–Sep		●	●		Sun
Silphium perfoliatum	Cup plant	N. Am.	2.0 m	Yellow	Jul–Sep			●		Sun
Silphium terebinthinaceum	Prairie dock	N. Am.	2.0 m	Yellow	Jul–Sep			●		Sun
Solidago gigantea	Giant goldenrod	N. Am.	1.5 m	Yellow	Aug–Sep		●	●		Sun
Solidago patula	Rough-leaved goldenrod	N. Am.	1.5 m	Yellow	Aug–Oct	●	●			Sun
Solidago ridellii	Riddell's goldenrod	N. Am.	1.0 m	Yellow	Aug–Sep	●	●			
Solidago rigida	Stiff goldenrod	N. Am.	1.0 m	Yellow	Jul–Oct			●	●	Sun
Solidago speciosa	Showy goldenrod	N. Am	1.0 m	Yellow	Jul–Oct			●	●	Sun
Stellaria palustris	Marsh stitchwort	N. Am.	30 cm	White	May–Jun		●	●		Sun
Symphytum caucasicum	Comfrey	Eur.	90 cm	Blue	May–Jun		●	●		Sun/shade Vigorous
Telekia speciosa		Eur.	2.0 m	Yellow	Jul–Sep		●	●		

HERBACEOUS PLANTS

Name	Common name	Origin	Height	Colour	Bloom time	Wet	Moist	Mesic	Dry	Notes
Thalictrum aquilegifolium	Meadow rue	Eur.	90 cm	Cream	May–Jul		●	●		Sun/partial shade
Thalictrum pubescens	Tall meadow rue	N. Am.	2.0 m	White	Jun–Jul		●	●		Sun/partial shade
Tradescantia ohiensis	Spiderwort	N. Am.	1.0 m	Blue	Apr–Jul		●	●		Sun
Trollius europaeus	Globe flower	Eur.	40 cm	Yellow	May–Jun		●	●		Sun/partial shade
Verbena hastata	Blue vervain	N. Am.	0.6 m	Purple	Jul–Oct	●	●			Sun
Vernonoa fasciculata	Ironweed	N. Am.	1.5 m	Purple	Jul–Sep	●	●	●		Sun
Veronica beccabunga	Brooklime	Eur.	0.3 m	Blue	Jun–Jul	●	●			Sun
Veronica longifolia	Longleaf speedwell	Eur.	1.0 m	Blue	Jul–Sep	●	●	●		Sun
Veronicastrum virginicum	Culver's root	N. Am.	1.0 m	White	Jun–Aug		●	●		Sun/partial shade
Viola pedata	Bird's-foot violet	N. Am.	0.1 m	Purple	Apr–Jun		●	●	●	Sun/shade
Zizia aurea	Golden Alexanders	N. Am.	1.0 m	Yellow	May–Jun		●	●		Sun/partial shade

GRASSES

Name	Common name	Origin	Height	Bloom	Wet time	Moist	Mesic	Dry	Notes
Andropogon gerardii	Big bluestem	N. Am.	1.5 m	Aug–Nov		●	●		Sun
Andropogon scoparius	Little bluestem	N. Am.	1.0 m	Aug–Oct			●	●	Sun
Arundo donax	Giant reed	Asia	3.0 m	Sep		●	●		Sun
Deschampsia cespitosa	Tufted hair grass	Eur. N. Am.	1.0 m	Jun–Jul		●	●		Sun/shade
Glyceria maxima	Reed sweet grass	Eur.	1.0 m	Jul	●	●			Sun
Glyceria occidentalis	Western manna grass	N. Am.	1.0 m	Jul	●	●			Sun
Juncus effusus	Soft rush	Eur. N. Am.	60 cm	Jun–Aug	●	●			Sun
Juncus inflexus	Hard rush	Eur.	60 cm	Jun–Aug	●	●			Sun
Miscanthus sinensis	Eulalia grass	Asia	2.0 m	Oct–Nov		●	●		Sun/light shade
Molinia caerulea	Purple moor grass	Eur.	60 cm			●	●		Sun
Panicum virgatum	Switch grass	N. Am.	1.2 m	Aug–Oct			●	●	Sun
Sorghastrum nutans	Indian grass	N. Am.	1.5 m	Aug–Oct			●	●	Sun
Sporobulus heterolepis	Prairie dropseed	N. Am.	1.0 m	Sep–Nov			●	●	Sun

SHRUBS

Name	Common name	Origin	Height	Colour	Bloom time	Wet	Moist	Mesic	Dry	Notes
Aronia arbutifolia	Chokeberry	N. Am.	M	White	May–Jun		●	●	●	
Aucuba japonica	Japanese laurel	Asia	M				●	●	●	Sun/shade
Calycanthus floridus	Sweet shrub	N. Am.	M	Purple	Apr–May			●		
Clethra alnifolia	Sweet pepper bush	N. Am.	L	White	Aug	●	●			
Cornus alternifolia	Pagoda dogwood	N. Am.	M						●	
Cornus sanguinea	Dogwood	Eur.	M	White	Apr	●	●	●		
Corylus americana	American hazelnut	N. Am.	L						●	
Corylus avellana	Hazel	Eur.	L				●	●		
Enkianthus campanulatus		Asia	L	Cream	Jun		●	●		
Fatsia japonica	Castor oil plant	Asia	L	White	Nov		●	●	●	Sun/shade
Frangula alnus	Alder buckthorn	Eur.	L			●	●			
Hamamelis virginiana	Witch hazel	N. Am.							●	
Hydrangea quercifolia	Oak leaf hydrangea	N. Am.	M	White	Aug–Sep		●	●		Sun/ partial shade
Ilex decidua	Possumhaw	N. Am.	M	Red berries	Jun		●		●	
Ilex glabra	Inkberry	N. Am.	M							
Ilex verticillata	Winterberry	N. Am.	M	Red berries	Jun		●	●		
Itea virginica	Virginia sweetspice	N. Am.	M	White	Jun		●	●		
Ledum palustre	Bog tea	N. Am. Eur.	M			●	●			

SHRUBS

Name	Common name	Origin	Height	Colour	Bloom time	Wet	Moist	Mesic	Dry	Notes
Mahonia aquifolium	Oregon grape	N. Am.	S	Yellow	Apr		●	●	●	
Physocarpus opulifolius	Ninebark	N. Am.	M	White	Jun			●		
Ribes nigrum	Blackcurrant	Eur.	M	Green	Apr	●	●	●		
Rubus odoratus		N. Am.	L	Pink	Jun–Jul		●	●		
Salix caprea	Goat willow	Eur.	L			●	●	●	●	
Salix cinerea	Sallow	Eur.	L			●	●			
Salix purpurea	Purple willow	Eur.	L			●	●	●		
Salix viminalis	Osier	Eur.	L			●	●			
Sambucus canadensis	Common elderberry	N. Am.	L	White	Jul			●	●	
Skimmia japonica	Skimmia	Asia	S	White	Apr			●	●	
Vaccinium uliginosum	Bog blueberry	N. Am.	S			●	●			
Viburnum dentatum	Arrow-wood	N. Am.	L	White	May–Jun			●		
Viburnum opulus	Guelder rose	Eur.	M	White	May–Jun					

TREES

Name	Common name	Origin	Height	Wet	Moist	Mesic	Dry
Acer circinatum	Vine maple	N. Am.	S		●	●	
Acer ginnala	Amur maple	Asia	M		●	●	
Acer rubrum	Red maple	N. Am.	L			●	
Aesculus octandra	Yellow buckeye	N. Am.	L			●	●
Alnus cordata	Italian alder	Eur.	M		●	●	
Alnus glutinosa	Common alder	Eur.	M	●	●	●	
Alnus incana	Speckled alder	Eur.	M	●	●	●	
Alnus rubra	Red alder	N. Am.	M	●	●	●	
Alnus serrulata	Tag alder	N. Am.	M	●	●		
Amelanchier spp.	Service berry	N. Am.	M			●	
Betula lenta	Cherry birch	N. Am.	M			●	
Betula nigra	River birch	N. Am.	M		●	●	●
Betula pubescens	Downy birch	Eur.	M	●	●		
Carpinus caroliniana	Ironwood	N. Am.	M		●	●	
Cercis canadensis	Redbud	N. Am.	M			●	●
Chionanthus virginicus	Fringe tree	N. Am.	M			●	
Fraxinus pennsylvanica	Green ash	N. Am.	L		●		
Liquidambar styraciflua	Sweet gum	N. Am.	L		●		
Nyssa sylvatica	Blackgum	N. Am.	L	●	●		
Populus tremula	Aspen	Eur.	M		●	●	
Prunus padus	Bird cherry	Eur.	S		●	●	
Quercus phellos	Willow oak	N. Am.	L			●	●
Salix alba	White willow	Eur.	L		●	●	
Salix fragilis	Crack willow	Eur	L	●			
Taxodium distichums	Swamp cypress	N. Am.	L	●	●	●	●

REFERENCES

Carter, T. and Rasmussen, T. 2005. Use of green roofs for ultra-urban stream restoration in the Georgia Piedmont (USA). In *Proceedings, Third North American Green Roof Conference: Greening Rooftops for Sustainable Cities*, Washington DC, 4–6 May 2005. Toronto: The Cardinal Group. 526–539.

City of Chicago. 2003. A Guide to Stormwater Best Management Practices: Chicago's Water Agenda. Available from http://egov.cityofchicago.org/webportal/COCWebPortal/COC_ATTACH/GuideToStormwaterBMPs.pdf Accessed 15 September 2006.

City of Portland Environmental Services. 2004. Stormwater Management Manual. Available from http://www.portlandonline.com/bes/index.cfm?c=35122 Accessed 15 September 2006.

Coffman, L. 2002. Low-impact development: an alternative stormwater management technology. In *Handbook of Water Sensitive Planning and Design*. Ed. R. L. France. Washington DC: Lewis Publishers.

Coffman, L. and Winogradoff, D. 2002. *Prince George's County Bioretention Manual*. Program and Planning Division, Dept of Environmental Resources, Prince George's County, Maryland.

Department for Transport. 2005. Road Casualties Great Britain: 2004 Annual Report. London: The Stationery Office.

Dunnett, N. and Kingsbury, N. 2003. *Planting Green Roofs and Living Walls*. Portland, Oregon: Timber Press.

Ferguson, B. 2002. Stormwater management and stormwater restoration. In *Handbook of Water Sensitive Planning and Design*. Ed. R. L. France. Washington DC: Lewis Publishers.

Federal Interagency Stream Restoration Working Group (FISRWG). 1998. Stream Corridor Restoration: Principles, Processes and Practices. USDA: Washington. Available from http://www.nrcs .usda.gov/technical/stream_restoration. Accessed 15 September 2006.

Hart, R. 1979. *Children's Experience of Place*. New York: Wiley.

Hitchmough, J. 2003. Herbaceous plant communities. In *The Dynamic Landscape: Ecology, Design and Management of Urban Naturalistic Vegetation*. Eds N. Dunnett and J. Hitchmough. London: Spon Press.

Kennedy, C. E. J. and Southwood, T. R. E. 1984. The number of species of insects associated with British trees: a re-analysis. *Journal of Animal Ecology* 53: 455–478.

Kennedy, M. 1997. The use and value of water. In *Designing Ecological Settlements*. Ed. M. Kennedy and D. Kennedy. Berlin: Dietrich Reimer Verlag.

Kohler, M., Schmidt, M., Grimme, F. W., Laar, M. and Gusmao, F. 2001. Urban water retention by green roofs in temperate and tropical climates. In *Proceedings of the 38th World Congress of the International Federation of Landscape Architects*, Singapore. Versailles: IFLA.

Liptan, T. 2002. Water gardens as stormwater infrastructure (Portland, Oregon). In *Handbook of Water Sensitive Planning and Design*. Ed. R. L. France. Washington DC: Lewis Publishers.

Littlewood, M. 2006. *Natural Swimming Ponds*. Atglen, PA: Schiffer Publishing.

Meiss, M. 1979. The climate of cities. In *Nature in Cities*. Ed. I. Laurie. Chichester: John Wiley & Sons.

Mentens, J., Raes, D. and Hermy, M. 2003. Effect of orientation on the water balance of green roofs. In *Proceedings, First North American Green Roof Conference: Greening Rooftops for Sustainable Cities*, Washington DC, May 2003. Toronto: The Cardinal Group. 363–371.

Moore, R. 1986. *Childhood's Domain: Play and Place in Child Development*. Beckenham: Croom-Helm.

Moran, A., Hunt, B. and Smith, J. 2005. Hydrologic and water quality performance from green roofs in Goldsboro and Raleigh, North Carolina. In *Proceedings, Third North American Green Roof Conference: Greening Rooftops for Sustainable Cities*, Washington DC, 4–6 May 2005. Toronto: The Cardinal Group. 512–525.

Mueller, A., France, R. and Steinitz, C. 2002. Aquifer recharge management model: evaluating the impacts of urban development on groundwater resources (Galilee, Israel). In *Handbook of Water Sensitive Planning and Design*. Ed. R. L. France. Washington DC: Lewis Publishers.

Owen, J. 1991. *The Ecology of a Garden: The First Fifteen Years*. Cambridge: Cambridge University Press.

Peck, S. P., Callaghan, C., Kuhn, M. E. and Bass, B. 1999. Greenbacks from greenroofs: Forging a new industry in Canada. Canada Mortgage and Housing Corporation.

Royal Society for the Prevention of Accidents. 2002. https://www.rospa.com/waterandleisuresafety/drownings /2002statistics.htm. Accessed 15 September 2006.

Smith, R., Warren, P. H., Thompson, K. and Gaston, K. 2005. Urban domestic gardens (VI): environmental correlates of invertebrate species richness, *Biodiversity and Conservation* online first.

Thayer, R. 1982. Public response to water-conserving landscapes. *Horticultural Science* 17: 562–565.

University of Wisconsin. 2003. *Rain Gardens: A How-to Manual for Homeowners*. Madison, WI: University of Wisconsin Extension Publications.

Wilkinson, D. M. 2001. Is local provenance important in habitat creation? *Journal of Applied Ecology* 38: 1371–1373.

Williams, P., Biggs, J., Corfield, A., Fox, G., Walker, D. and Whitfield, M. 1997. Designing new ponds for wildlife. *British Wildlife* 8: 137–150.